AF846082

Unravelling the Fukushima Disaster

The Fukushima disaster continues to appear in national newspapers when there is another leakage of radiation-contaminated water, evacuation designations are changed, or major compensation issues arise, and so remains far from over. However, after five years, attention and research towards the disaster seems to have waned despite the extent and significance of the disaster that remains.

The aftermath of Fukushima exposed a number of shortcomings in nuclear energy policy and disaster preparedness. This book gives an account of the municipal responses, citizen's responses, and coping attempts, before, during, and after the Fukushima crisis. It focuses on the background of the Fukushima disaster, from the Tohoku earthquake to diffusion on radioactive material and risk miscommunication. It explores the processes and politics of radiation contamination, and the conditions and challenges that the disaster evacuees have faced, reflecting on the evacuation process, evacuation zoning, and hope in a post-Fukushima environment.

The book will be of great interest to students and scholars of disaster management studies and nuclear policy.

Mitsuo Yamakawa is Professor of Economic Geography at Teikyo University and Extraordinary Professor of Fukushima Future Center for Regional Revitalization (FURE) at Fukushima University.

Daisaku Yamamoto is Associate Professor of Geography and Asian Studies at Colgate University, New York, USA.

Routledge Studies in Hazards, Disaster Risk and Climate Change

Series editor: Ilan Kelman, Reader in Risk, Resilience and Global Health at the Institute for Risk and Disaster Reduction (IRDR) and the Institute for Global Health (IGH), University College London (UCL).

This series provides a forum for original and vibrant research. It offers contributions from each of these communities as well as innovative titles that examine the links between hazards, disasters and climate change, to bring these schools of thought closer together. This series promotes interdisciplinary scholarly work that is empirically and theoretically informed, with titles reflecting the wealth of research being undertaken in these diverse and exciting fields.

Cultures and Disasters
Understanding cultural framings in disaster risk reduction
Edited by Fred Krüger, Greg Bankoff, Terry Cannon, Benedikt Orlowski and E. Lisa F. Schipper

Recovery from Disasters
Ian Davis and David Alexander

Men, Masculinities and Disaster
Edited by Elaine Enarson and Bob Pease

Unravelling the Fukushima Disaster
Edited by Mitsuo Yamakawa and Daisaku Yamamoto

Unravelling the Fukushima Disaster

Edited by Mitsuo Yamakawa and Daisaku Yamamoto

Routledge
Taylor & Francis Group
LONDON AND NEW YORK

First published 2017
by Routledge
2 Park Square, Milton Park, Abingdon, Oxon OX14 4RN

and by Routledge
711 Third Avenue, New York, NY 10017

Routledge is an imprint of the Taylor & Francis Group, an informa business

© 2017 selection and editorial matter, Mitsuo Yamakawa and Daisaku Yamamoto; individual chapters, the contributors

The right of the editors to be identified as the authors of the editorial material, and of the authors for their individual chapters, has been asserted in accordance with sections 77 and 78 of the Copyright, Designs and Patents Act 1988.

All rights reserved. No part of this book may be reprinted or reproduced or utilised in any form or by any electronic, mechanical, or other means, now known or hereafter invented, including photocopying and recording, or in any information storage or retrieval system, without permission in writing from the publishers.

Trademark notice: Product or corporate names may be trademarks or registered trademarks, and are used only for identification and explanation without intent to infringe.

British Library Cataloguing in Publication Data
A catalogue record for this book is available from the British Library

Library of Congress Cataloging in Publication Data
Names: Yamakawa, Mitsuo, 1947- editor. | Yamamoto, Daisaku.
Title: Unravelling the Fukushima disaster / edited by Mitsuo Yamakawa and Daisaku Yamamoto.
Description: New York: Routledge, [2017] | Series: Routledge studies in hazards, disaster risk and climate change | Includes bibliographical references and index.
Identifiers: LCCN 2016023543| ISBN 9781138193819 (hardback) | ISBN 9781315639130 (ebook)
Subjects: LCSH: Tohoku Earthquake and Tsunami, Japan, 2011. | Earthquake damage–Japan. | Nuclear power plants–Accidents–Japan. | Nuclear power plants–Natural disaster effects. | Natural disasters–Environmental aspects.
Classification: LCC HV600 2011 .T64 U567 2017 | DDC 363.34/950952–dc23
LC record available at https://lccn.loc.gov/2016023543

ISBN: 978-1-138-19381-9 (hbk)
ISBN: 978-1-315-63913-0 (ebk)

Typeset in Times New Roman
by Sunrise Setting Ltd, Brixham, UK

Contents

Figures

Tables

Contributors

Sasha Davis is an Assistant Professor of Geography and Environmental Science at the University of Hawaii-Hilo. His studies have focused on the environmental, cultural, and political aspects of redeveloping landscapes contaminated by nuclear-weapons testing and other military activities. His previous research has taken place in Bikini Atoll, Okinawa, Guam, Hawaii, and Puerto Rico. His research has been published in the *Annals of the Association of American Geographers, Environment and Planning D*, and the *Professional Geographer*. More recently, his research on environmental issues, social movements, and geopolitics has been published as a book titled *Empires' Edge: Militarization, Resistance, and Transcending Hegemony in the Pacific.*

David W. Edgington is Professor of Geography at the University of British Columbia, where he teaches courses on the geography of Asia and the Pacific Rim. He is author of *Reconstructing Kobe*, published by the UBC Press in 2010. His current research themes include reconstruction of Tohoku, Japan, after the 3.11 triple disaster, the geography of Japanese business investment in the Greater China region, and urban governance issues in Japanese cities.

Noritsugu Fujimoto is Professor of the Faculty of Regional Development Studies, Toyo University. He holds a PhD in economics from Kyushu University, Fukuoka. His main research areas include regional economic development, urban systems, and office-location dynamics.

Jessica Hayes-Conroy is an Assistant Professor of Women's Studies at Hobart and William Smith Colleges. Her work has focused on the intersection of bodies, health, and the environment, especially studies on food and nutrition. Her previous research has taken her to Vieques, Puerto Rico, Nova Scotia, Canada, and Medellin, Colombia, among other places. Her work has been published in *Gender, Place and Culture, Environment and Planning A*, and *Emotion, Space and Society*. She has recently published a book on school garden and cooking programs entitled *Savoring Alternative Food: School Gardens, Healthy Eating and Visceral Difference*.

Naoko Horikawa is a Project Researcher in the Fukushima Future Centre for Regional Revitalization (FURE) at Fukushima University. She received her PhD in Social Anthropology from the University of Hull, UK. Her current research investigates the lives of evacuees following the disaster at the Fukushima nuclear-power plant in 2011. In her dissertation "New Lives in the Ancestral Homeland: Return migrants from South America to mainland Japan and Okinawa" (University of Hull, 2012) she comparatively analyzed identity-formation processes among Nikkeijin and Okinawan returnees to Japan.

Hiroyuki Kaneko holds a PhD in Human Sciences and is Postdoctoral Fellow of Japan Society for the Promotion of Science (JSPS). His publications include "Village Conflict and Distribution of Flood Disaster" (in Ueda, ed., 2015, *Coping with Disasters: Communities and Families*, in Japanese), and "Settlers in Rivers and Their Disaster Response" (in Torigoe, ed., 2013, *Use and Destruction of Nature*, in Japanese).

Kencho Kawatsu is Professor of the Faculty of Symbiotic Systems Science, and the manager of the Division of Measures against Radioactive Contamination in the Fukushima Future Center for Regional Revitalization (FURE) at Fukushima University. He worked in the area of nuclear politics and environmental policy at Fukushima Prefectural Government over many years.

Kyo Kitayama is Research Fellow in the Division of Measures against Radioactive Contamination in the Fukushima Future Center for Regional Revitalization (FURE) at Fukushima University, and he has a PhD in agricultural studies. His research interests include the atmospheric behavior of radiocesium, atmospheric chemistry in aerosols, and precipitation.

Katsumi Nakai is President of Fukushima University and Professor of the Faculty of Administration and Social Sciences. His main research areas include legal science, public administration, and environmental law. He is a member of the Japan Public Law Association and the Japan Association for Environmental Law and Policy.

Yosuke Nakamura is Associate Professor of the Faculty of Human Development and Culture at Fukushima University, and he holds a PhD in physical geography from Kyoto University. He has studied tectonic geomorphology and natural disasters and elucidated the recurrence intervals of active faults and the mechanism of large landslides and flooding disasters in East Asia. His current activities include education for disaster prevention for students and children in the Fukushima area.

Takashi Oda is Associate Professor in the Center for Disaster Education and Recovery Assistance at Miyagi University of Education, and he holds PhD in environmental studies and an MSc in earth science from Tohoku University. As a geographer, his recent interests involve the spatial understanding of disasters and education for disaster-risk reduction, and since March 2011 he has been actively engaged in the academic and civic communities' efforts to examine the impacts of the disaster and to assist the recovery and rebuilding in the afflicted areas. Prior to his current appointment, he worked at the Center for

Simulation Sciences, Ochanomizu University in Tokyo and was a Postdoctoral Research Fellow of the Japan Society for the Promotion of Science (JSPS).

Kenji Ohse is Associate Professor in the Division of Measures against Radioactive Contamination in Fukushima Future Center for Regional Revitalization (FURE) at Fukushima University. He specializes in environmental soil chemistry and radioecology. He is currently studying the behavior of radionuclides in agro ecosystems.

Mitsuo Yamakawa is Professor of Economic Geography at Teikyo University and Extraordinary Professor of Fukushima Future Center for Regional Revitalization (FURE) at Fukushima University. He is also a Regular Member of the Science Council of Japan. His recent books include *Economic Geography on Revitalization from Fukushima Nuclear Disaster* and *Japanese Economy and Regional Structure* (both in Japanese). He is the Project Leader of *Establishing Academic Framework of Earthquake Disaster Reconstruction Experiencing Great East Japan Earthquake*, funded by Grant-in-Aid Scientific Research (Category S) of Japan.

Daisaku Yamamoto is Associate Professor of Geography and Asian Studies at Colgate University, Hamilton, New York. He holds a PhD in geography from the University of Minnesota. His recent work focuses on community resilience, regional inequality, and uneven development. He is currently working on the project to examine the socio-economic effects of nuclear decommissioning on local communities in the US and Japan.

Preface and Acknowledgments

The earthquake on March 11, 2011 and the subsequent tsunami and nuclear accident took thousands of lives and changed forever the lives of tens of thousands of people who lived along the coast of the Tohoku region of Japan. Immediately after the earthquake and tsunami, the areas swept by their mighty and ruthless force resembled the landscape after the air raids of World War II that burned and leveled the cities of Japan: buildings destroyed without a trace and washed-up wreckage dotting the scene. Five years after the disaster, signs of recovery and reconstruction are evident in many of the devastated areas. However, in the towns near Tokyo Electric Power Company's (TEPCO), Fukushima Daiichi Nuclear Power Plant (NPP), you can still see buildings destroyed by the earthquake and tsunami as if time came to a stop on March 11. One cannot help feeling a sense of profound human futility in the face of a nuclear accident releasing a significant amount of radioactive materials across the landscape.

Indeed, while the damage from the earthquake and tsunami was enormous, it is environmental contamination by radioactive materials that has been the greatest source of affliction for many residents of Fukushima Prefecture. This is especially the case for inhabitants of the communities around the NPP who were ordered to evacuate, but it is also the case for those living outside mandatory evacuation zones yet still within higher-than normal radiation areas. These individuals have had their families torn apart, their livelihoods suspended or altered, their health severely weakened, and their children's education interrupted. Many of the evacuated areas have been gradually labeled as "safe to return" after government-led decontamination projects; yet the process of returning has been decidedly slow. Farmers and businesses still suffer from tangible damage and the stigma associated with radioactive contamination.

The human casualties, environmental damage, and socio-economic distress all represent what we have lost in this historic disaster. Yet they are not all of what we lost. The poor responses that followed the earthquake, tsunami and subsequent nuclear-power accident brought heavy criticism and condemnation not only for politicians, the government-corporate nexus and the media but also for the academic community, even for scientific knowledge itself. This eroding faith in authority, experts, and science has been encapsulated in and exacerbated by the plethora of new phrases and terms that filled the media after March 11.

Prominent among these terms are "unexpected," "no immediate effects," and "interim limits." For example, the notion that a nuclear disaster is "unexpected" means that it is assumed to be beyond the range of the possible. It was on the foundation of this assumption—and also a flagrant over-confidence in the superiority of Japanese manufacturing and engineering—that the paradigm of NPP safety was erected. "No immediate effects" also carries in its train a whole host of troublesome baggage. This term emerges out of a politically motivated attempt to quell the anger and agitation of victims and the general public, but it can only be uttered through the willful disregard of substantial scientific evidence concerning the actual state of air radiation dose levels in the afflicted areas and the health effects of radiation exposure. Simply relying on data from the International Commission on Radiological Protection (ICRP)—a group whose voice is among the choir chanting the refrain of nuclear safety—to assess the health effects of low-dose-radiation exposure on children is by no means sufficient to dispel anxiety for parents and the public. Finally, the phrase "interim" has been used as a means of altering the limits for radiation in food for a certain but undefined period of time, and it is this declaration of an interim of exception that has been the source of much disbelief among victims and the public, leading to the issue of stigmatization, or reputational damage (*fuhyo higai*).

The appearance of these three paradigmatic examples of nuclear-disaster discourse and procedure are an exemplary illustration of the way in which the formerly paired terms of *anzen* and *anshin* ("safety and reassurance") have been clearly divided, such that, even when objective criteria are used to proclaim "safety," we are no longer able to feel "reassured." Even so, various government agencies, technocrats, and public officials have continued on various occasions, such as public hearings and press conferences, to "ask for the understanding and cooperation of the citizens," a uniquely Japanese bureaucratic expression that tactfully asserts a one-way flow of correct, expert knowledge and implicitly belittles the public for its ignorance. We must demystify the so-called expert knowledge and power associated with it. To do so inevitably demands a critical reflection on knowledge that *we* produce. In a country where the development of academic institutions has been tightly linked with the process of modernization and industrialization, such critical reflection and demystification of knowledge is not easily undertaken.

We would also like to emphasize, nevertheless, that we still believe in the importance of the knowledge that we produce—in this case, for its practical usefulness in the rebuilding and revitalization of the livelihoods of the people who have been torn away from their familiar environment and communities, rather than focusing exclusively on criticizing the developmental state and other power asymmetries. It is in this context that a team of researchers, headed by one of the editors (Yamakawa), at the Fukushima University decided to pursue a multi-year transdisciplinary project, supported by the *KAKENHI* Grant-in-Science Project, on recovery, reconstruction, and redevelopment from the Great East Japan Earthquake Disaster, the term we use in this book to refer inclusively to the triple disasters in Japan on March 11, 2011 and afterwards: the earthquake, the tsunami, and the nuclear accident. Many of the contributors to this book and another book that is being published concurrently (Yamakawa and Yamamoto, 2017) are members of this research project. Other contributors come from various research institutions in and outside of Japan and have been conducting research on Fukushima over several years. Despite the wide range of institutional and disciplinary backgrounds of these authors, they are united by a shared concern and desire to contribute to the rebuilding and revitalization of the livelihoods of those who were affected by the historic disaster.

One of the editors (Yamakawa) was working as an economic geographer in the School of Economics and Management at Fukushima University, located in the prefectural capital city, at the time of disaster. As the only national university in the prefecture, it felt an urgent need to respond to multiple challenges on issues ranging from direct damage caused by the earthquake and tsunami to radioactive contamination, support for more than 100,000 evacuees, and assistance for local communities and industries. In April 2011, the university established the Fukushima Future Center for Regional Revitalization (FURE) as its outreach institute to address these issues. Yamakawa became the first director of the center upon the recommendation of the president of the university, Osamu Nittono, and of the vice president, Akira Watanabe. He would like to express his profound gratitude to these two individuals for providing such challenging yet rewarding opportunities.

Seiichi Chiaki, the head of the center office, and Yutaka Yamazaki, the associate head, provided enormous support in developing the center's organizational architecture and hiring staff researchers. The center started out with four divisions: Children and Youth Support Division, Reconstruction Planning Support Division, Energy Environment Division, and Planning and Coordination Division. The following members of the center played pivotal roles in research activities for reconstruction and in the publication of this book: Toshio Hatsuzawa (director of the center); Hiroyasu Shioya, Katsuhiko Yamaguchi, Ryota Koyama, and Tomotaka Mori (division heads); and Yasumichi Nakai, Fuminori Tamba, Naoaki Shibasaki, Kencho Kawatsu, Michio Sato, Noriko Yoshinaga, and Atsushi Igarashi.

Some of the chapters of this book are based on research supported by the five-year KAKENHI Grant-in-Aid for Scientific Research (Category S), 2013–17, through the Japan Society for the Promotion of Science. Yamakawa applied for this grant in March 2013, before his retirement from Fukushima University, and the grant was subsequently awarded in June that year. The proposal for the grant was drafted by the members of the FURE, including Yosuke Nakamura, Itsuki Yoshida, and Akira Takagi. The research project has been carried out by a number of professors and instructors of Fukushima University (unless otherwise noted), including Akira Takagi, Akihiko Sato, Toshio Hatsuzawa, Satoru Mimura, Hideki Ishii, Hiroshi Kainuma, Katsumi Nakai (President of Fukushima University), Kenji Ohse, Yoshio Ohira, Kyo Kitayama, Noritsugu Fujimoto, Koichiro Matsui (Teikyo University), and Kosei Yamada (Teikyo University). Furthermore, Masayuki Seto (doctoral researcher), hired as a project manager, oversaw and managed the project while contributing to the research itself. Without his dedication, the KAKENHI project would not have progressed. We thank all the individuals who engaged in this research project, which provided an important basis for the book.

The members of the FURE and of the KAKENHI project have been advancing their research, around the keyword *shienchi* ("support knowledge"), through intensive and numerous interactions with local governments, organizations, and residents in disaster-afflicted areas of Fukushima. In particular, the research would not have been possible without the generous cooperation, wisdom, and patience of Yuko Endo (Mayor of Kawauchi Village), Katsunobu Sakurai (Mayor of Minami Soma City), Tamotsu Baba (Mayor of Namie Town), Norio Kanno (Mayor of Iitate Village), officers of these municipalities, officers of the prefectural divisions, specialized personnel of not-for-profit organizations, and, above all, those evacuees who have been forced to live away from their homes. It is our hope that the center, the research project, and this book will help their efforts to rebuild their livelihoods.

Yamakawa, as a council member of the Japan Society for the Promotion of Science, has been involved in a number of advocacy activities related to the recovery and reconstruction of nuclear-disaster-afflicted regions since March 2011. In particular, he played central roles in proposing the "Urgent recommendation on the development of inspection systems as a counter measure to the stigmatization of food and agriculture as the result of the nuclear disaster" (September 6, 2013), and "Recommendations on the reconstruction of livelihood and housing for long-term evacuees as the result of the TEPCO Fukushima Daiichi Nuclear Power Plant accident" (September 30, 2014). To develop these proposals, Ryota Koyama and Fuminori Tamba provided substantial assistance.

The editors knew each other prior to the disaster and met again at the annual conference of the Japan Association of Economic Geographers in Tokushima, Japan in the spring of 2012. However, the idea of publishing an English-language book (which turned out to be two books) did not arise until one of the editors (Yamamoto) and Noritsugu Fujimoto co-organized a series of sessions titled *Fukushima: Three Years Later* at the Annual Meeting of the Association of American Geographers in Tampa, Florida in April 2014. The sessions offered opportunities for Japanese researchers working on the KAKENHI and other projects on Fukushima and those researchers outside of Japan who have paid close attention to the nuclear disaster in Fukushima to exchange their knowledge and experiences. Many of the chapters of this book were developed based on the presentations at the sessions. We thank all those who participated in the sessions as presenters, commentators, and audience, including Thomas Feldoff, Laura Beltz Imaoka, Noriko Iwai, Satoru Masuda, Robert Mason, Marie Augendre, Yuta Hirai, Nate Pickett, Peter Klepeis, and Eric Sheppard. After the AAG session, we had a few occasions at Colgate University in which we had opportunities to learn further about Fukushima. Hiroyuki Torigoe, Kazunori Matsumura, and Arthur Binard were among some of the most inspiring speakers who indirectly nourished our work.

We also thank Jay Bolthouse, who translated some of the manuscripts originally written in Japanese. His work went beyond mere translation of texts to become the translation of knowledge through his critical attention to the quality of arguments, logical flows, and data presentation. Furthermore, we thank Bill Meyer (Colgate University), who provided much help in revising many of the manuscripts and offered numbers of useful comments, and Yamamoto's undergraduate students Angelica Greco and Samantha Trovillion, who offered their assistance in editing and proofreading the manuscripts. Working with Melissa Heller, Julia Feikens, Mallory Hart, Jessica Li, and Madelin Horner on the research on nuclear decommissioning also informed this project in important ways.

We extend our appreciation to Colgate University, its Research Council, and several of its departments and programs that offered financial support for part of the project, including the AAG sessions and the hosting of the symposium on Fukushima in April 2014. We would like to acknowledge the members of the editorial and production staff at Routledge, including Faye Leerink, Priscilla Corbett, and Laura Johnson, for their professionalism, patience, and unfailing helpfulness, especially working with someone who had limited experience in publishing books in English.

Finally, we would like to express our greatest appreciation and deepest indebtedness to Ryoko Yamakawa and Yumiko Yamamoto for all their support and patience during the much-too-long process of the project leading up to this book.

Reference

Yamakawa, Mitsuo and Daisaku Yamamoto. eds. 2017. *Rebuilding Fukushima*. London: Routledge.

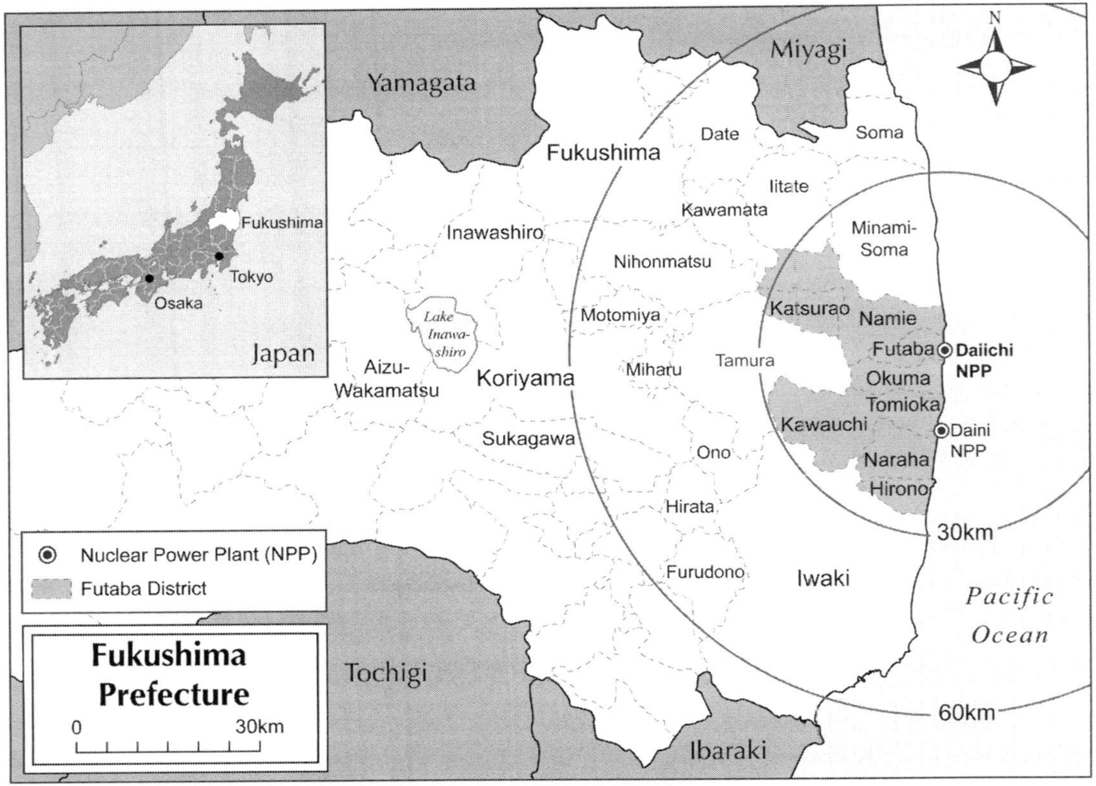

Map of Fukushima Prefecture
Created by Masayuki Seto and Daisaku Yamamoto

Introduction

Mitsuo Yamakawa, Katsumi Nakai, and Daisaku Yamamoto

A large number of publications, scholarly and otherwise, on the Fukushima nuclear disaster have appeared and continue to appear in Japan. "Fukushima" returns to the front page of national newspapers when there is another leakage of radiation-contaminated water, evacuation designations are changed, or major compensation issues arise, but for the most part we can pass days without seeing or thinking anything about the nuclear disaster today. After all, five years is a long time for any disaster to stay in the memory of those who were not directly affected by it. Yet, you realize quickly, if you pay even a slight attention to—or better yet, visit—the nuclear disaster-afflicted areas of Fukushima, that the disaster is far from over. It concerns us, then, that given the extent and the significance of the disaster, interest in Fukushima seems to be waning even among researchers in Japan.

Social-scientific books and articles in English on Fukushima emerged rapidly after the disaster, but most of these books focus on why the accident happened (Carpenter, 2012; Hindmarsh, 2013), how the disaster, i.e. meltdown, unfolded (Lochbaum *et al.* 2014; Nadesan, 2013; Willacy, 2013), how Japan and its media reacted (Elliott, 2012; Kingston, 2012), what political and institutional changes will and should happen in Japan (Hindmarsh, 2013; Kingston, 2012; Samuels, 2013), and how other countries reacted and should learn from the disaster (Elliott, 2012; Lochbaum *et al.*, 2014). For example, Carpenter (2012) explores the political economic apparatus that gave rise to nuclear energy development in Japan, inherent problems of the regulatory mechanisms of supervision, and institutional rigidities that shape the growing public opposition to nuclear power, essentially explaining why this historical disaster was bound to happen. Willacy (2013) provides an excellent account of how the successive meltdown of the reactors took place as the result of the earthquake, tsunami, and human errors in the aftermath of March 11, 2011, incorporating vivid individual stories of victims, plant workers, and officers. Other authors such as Samuels (2013) and Hindmarsh (2013) address the questions of what the Japanese and the state should learn from the disaster, how to reform their energy policies and governance structure, and how (un)likely such reforms are.

Unfortunately, but perhaps not surprisingly, there have been relatively few English-language scholarly works that examine what has been happening in Fukushima since the immediate crisis situation in 2011 and what is actually being done in order to secure and revive the livelihoods of its people. In part this is probably because the "mundane" but highly complex issues associated with the reconstruction of livelihoods are far less dramatic than the scenes of explosions, destruction, and ghost towns, and do not permit straightforward, unequivocal critiques of such entities as the state, TEPCO, and the media, which can be more easily executed without being in Fukushima on a regular basis. As a result, little is known outside of Japan about the extent, magnitude, and complexity of radioactive contamination, the lives of residents and evacuees who face difficult decisions about whether to stay or leave, and the farmers and businesses suffering tangible losses and the stigma associated with radioactive contamination. This book and *Rebuilding Fukushima* (forthcoming) address these issues that manifest themselves in the vast "gray zones" of Fukushima, i.e. outside of areas that are unlikely to be inhabitable for the foreseeable future but still within areas where above-normal radiation levels are a concern.

Unfolding of the nuclear disaster

The earthquake that triggered the two subsequent calamities, the tsunami and the nuclear accident, is more specifically referred to as the 2011 Off the Pacific Coast of Tohoku Earthquake. Nakamura (Chapter 1) describes the basic outlines of this earthquake, which is at the core of the complex nuclear disaster that followed. The earthquake, which occurred at the plate boundary between the Pacific Plate and the North American Plate on March 11, 2011, was followed by thousands of aftershocks. The chapter shows that seismic activity in Eastern Japan after the main earthquake falls into at least four distinct types, each with its own geophysical dynamics, and that aftershocks and induced earthquakes are expected to continue at least for the next ten years. This implies that these earthquakes may potentially affect the containment processes of the nuclear accident and the recovery efforts. The chapter also reminds us that most of the nuclear power stations in Japan are located in seismically "shaky" zones.

Significant quantities of radioactive materials were released as the result of the Fukushima Daiichi Nuclear Power Plant accident, thereby seriously damaging the socio-ecological systems in Fukushima and beyond. Kawatsu, Ohse, and Kitayama (Chapter 2) trace the "invisible" physio-spatial processes of diffusion and deposition of radioactive materials, particularly radionuclides such as iodine and cesium. The behavior of these radionuclides must be examined at multiple dimensions because its dynamics differ considerably between atmospheric and terrestrial environments, between various spatial scales, between different soil types, and even between different crops. The authors conclude that we now have a relatively good understanding of the extent and mechanisms of the transfer of radionuclides in atmospheric and terrestrial environments, at least for the early period of the nuclear accident. Nevertheless, they say that radionuclides, once transferred to food, began a whole new set of socio-political dynamics. Despite rigorous monitoring efforts and actions to limit the circulation of radioactively contaminated food in the market, the stigmatization of food produced in Fukushima has become a major concern.

The first evacuation order was issued by the national government on the night of March 11. At that time, areas within 2 km of the plant were designated as evacuation zones. As the accident grew increasingly severe, the area designated for evacuation was extended to 5 km, to 10 km and, by the evening of March 12, to 20 km from the site. In reality, however, the evacuation process was anything but orderly due to the poor handling of the situation by the government and TEPCO. For instance, the municipal office of Namie Town, less than 10 km from the nuclear plant, did not have any direct communication with TEPCO about the situation at the plant, nor did it receive any evacuation order directly from the national or prefectural government at the time of the first hydrogen explosion at the plant. Subsequently, the mayor issued his own evacuation order based on the information broadcast by the news media. First he urged the residents to move outside the 10 km radius of the plant (6 am on March 12) and then outside the 20 km radius (11 am on the same day). He recommended that they relocate to the Tsushima district of Namie Town, which was relatively far from the coast, using charter buses or their personal vehicles. However, it was later determined that external radiation exposure dose levels were high due to the direction of the plume. As is now widely known, SPEEDI system forecasts of the radiation plume went unutilized (finally becoming available to the public on March 23, 2011), and crucial evacuation orders went unissued for Namie Town, resulting in radiation exposure.

The continuous alteration of evacuation zones caused severe problems even for those who were outside of the forced evacuation zones, due to poor risk communication. Oda (Chapter 3) takes up the problem of stigmatization that started immediately following the onset of the nuclear accident, with a focus on the risk of *mis*communication observed in Iwaki City, Fukushima Prefecture. In particular, he suggests that poor communication of geographic information by mass media created a serious stigma, or "reputational damage" (*fuhyo higai*), that subsequently isolated the residents of Iwaki City and Minami-Soma City, located south and north of the power plant respectively, from the supply of essential goods and services. In short, grossly generalized information about areas of high radiation risks gave the impression that some areas of these cities that were far from the source of the disaster were highly dangerous.

Lives in suspension: nuclear disaster evacuees

Unlike earthquakes and tsunamis, nuclear disasters make it impossible for residents to return to towns and villages for a substantial amount of time, and this slows down and complicates unified efforts towards reconstruction and restoration. After multiple revisions of evacuation designations in the first year of the accident, afflicted areas were classified into three classes based on annual cumulative radiation dose: "evacuation lift preparation areas" (areas with an annual cumulative dose under 20 mSv/year), "limited residence areas" (annual cumulative dose ranging between 20–50 mSv/year), and "difficult to return areas" (annual cumulative dose exceeding 50 mSv/year). As a result of this re-designation,[1] significant portions of Futaba District, which includes eight municipalities, became "difficult to return areas," in particular Okuma Town and Futaba Town, where 90% of the residences had been evacuated.[2] By the fall of 2011, it was estimated that nearly 49,000 residents of Futaba District had been evacuated from their homes.

Yamakawa (Chapter 4) focuses on the former residents of Futaba District who were evacuated outside of the town. The chapter provides critical insights into the conditions for and perceptions among the Futaba evacuees in the first six months of the disaster, based on a comprehensive survey conducted between September and October 2011. The survey reveals a number of important characteristics of the nuclear-disaster evacuation, including dispersed destinations, variegated housing types, distrust of authority, differences in views and opinions across age and gender groups, and both the hope and despair of return. The results of the survey provide a number of vital policy implications to support the evacuees whose lives have been in prolonged suspension. It is imperative, Yamakawa argues, to respect and facilitate individual differences among evacuees in terms of their feelings and preferences for their future options.

Nearly four years after the nuclear accident, at the end of 2014, approximately 100,000 people from Fukushima were estimated still to be in the state of evacuation. Of those evacuees, some 20,000 were considered "voluntary evacuees," a controversial term that refers to those who relocated from their homes by processes not demanded or directly assisted by governments. Since this voluntary evacuation was not planned, the decision to head towards a certain destination was decided in nearly all cases by chance circumstances. It is due to the random, rather than planned, nature of this evacuation that approximately half of all evacuees from the Futaba district have had to change their evacuation destination more than five times. Voluntary evacuation has resulted in a wide scattering of evacuation destinations and has led to a situation where many evacuees are living in dispersed forms of housing, such as government-subsidized rentals units and other private-sector rentals, rather than more typical post-disaster forms of concentrated housing, such as temporary shelters. Indeed, over half of evacuees are living in dispersed rental housing, a first in the history of disasters and disaster housing in Japan; and, despite bringing some benefits, it has been partially responsible for furthering the spatial divisions that have torn families and communities apart.

One of the main causes of the division of families and communities has been the fundamental difference in perspectives through which different social groups evaluate the risks of radiation exposure. Concern over radiation exposure is generally higher among women than men, and this gender gap in perception is even more strongly in effect for women who are caring for children. Evidenced here, then, is the strong will of a mother to protect her children from the damaging health effects of radiation exposure. This difference in perception and practices is evident in the large number of husbands and fathers who remain in Fukushima and the large number of wives, mothers, and children who are leading separate lives outside the prefecture. Horikawa (Chapter 5) focuses on the voluntary evacuees, many of whom are mothers who fear the potential effects of radiation on their children (but with some important exceptions). They often face institutional obstacles because of their "voluntary" of status. Nevertheless, their concerns and actions must be viewed in the context of scientific uncertainty over the effects of long-term low-level radiation (such as 1 to 20 mSv/year) doses on the body. Accordingly, views on whether or not to pursue voluntary evacuation away from these lower radiation dose areas vary within and between families, relatives, schoolmates, and communities and have been a key source of conflict. Horikawa explores their experiences, thoughts and feelings, and coping strategies, which often escape media attention, through ethnographic narratives.

Living in a radiation-contaminated environment

Perhaps the most distinct challenge of the nuclear disaster is how to deal with radioactive contamination of living spaces. The nuclear accident at the Fukushima Daiichi Nuclear Power Plant resulted in radioactive contamination in the eastern areas of Fukushima Prefecture and other neighboring areas. The Cabinet decision permitted only persons living within a 20 km radius of the plant and around a limited number of major "hotspots" to receive compensation from TEPCO. Outside these official evacuation zones, there are vast areas of radiation levels greater than 5 mSv/year and additional highly localized "hotspots" identified by some nongovernmental organizations. In fact, if the standards for evacuation based on the Chernobyl nuclear accident were applied to Fukushima, two thirds of Fukushima Prefecture, spanning much of Hama-Dori (coastal area) and Naka-Dori (inland area), would be authorized as official evacuation zones. In other words, while extremely high radiation zones are obviously off limits for residents for the foreseeable future, a large number of residents in Fukushima have been living in a higher-than-normal radiation environment.

Immediately after the contamination of the Fukushima landscape as a result of the deposition of radioactive materials, the question of how to pursue the decontamination of land and buildings became a central and pressing one. Radioactive materials are like particles of dust and are deposited onto buildings, roads, or other surfaces. If the environment is to be decontaminated (which more accurately means the transfer of radioactive materials from one place to another), buildings and roads must be washed with pressured water, the surface (which is the most nutrient rich part) of soil must be removed, and leaves and branches must be disposed of.

Chapters 6 through 9 address the issues of living in and dealing with the radiation-contaminated environment in Fukushima from different perspectives. Edgington (Chapter 6) takes up the "cavernous disconnect" that has emerged between the Tokyo technocrats who argue that standards regarding evacuation, decontamination, and return should be calculated in terms of a quantifiable risk–benefits analysis and the residents and not-for-profit organizations (NPOs) of Fukushima who demand the full evacuation of contaminated areas, or at least of their more vulnerable residents, on ethical and humanitarian grounds. Drawing on the literature on risk analysis, the chapter examines questions surrounding how the government has approached radiation risks since March 2011, how the decontamination programs of local governments have been carried out, and what factors have shaped the risk perceptions of citizens and NPOs.

Fujimoto (Chapter 7) examines the issues of decontamination from a political-economy perspective. He argues that the evacuation zones of the Fukushima nuclear accident are determined primarily by fiscal constraints of the central and local governments and by their political motivations to obscure the extent of the damage from the nuclear accident. The governments, in partnership with TEPCO, are underplaying the possible public-health risks in order to avoid issuing the astronomical number of bonds that would be required to pay for additional evacuation compensation. Instead the governments promote "decontamination intensive" policies, which effectively protect the interests of existing stakeholders, such as the construction and electric power industries, and preserve the old industrial structure and the spatial economic hierarchy of cities.

While state-supported decontamination efforts have been under way, however problematic they may be, it remains a fact that many residents of Fukushima continue to live in the "gray zones." As with any natural or human-induced disaster, the effects of the Fukushima nuclear disaster have been experienced differently based on a myriad of socio-spatial factors, including age, gender, occupation, residence, and everyday spatial behaviors. Davis and Hayes-Conroy (Chapter 8) examine not only how people are affected differently by the disaster, but also how people's social identities and varied relationships to the natural environment can affect the decontamination, restoration, and redevelopment process. To illustrate this point, the chapter draws information from three sources: interviews conducted in 2013 with Fukushima residents about daily landscape interaction and food contamination; perspectives from geographic research traditions such as feminist geography and political ecology; and previous research undertaken in radioactive environments in the Marshall Islands. Davis and Hayes-Conroy argue that variable experiences of contamination will play an important role in determining the acceptability of future redevelopment policies and programs in Fukushima.

Coming from the intellectual tradition of Japanese environmental sociology and folk-culture studies, Kaneko (Chapter 9) provides us with critical insights into how people view, experience, and cope with the radiation-contaminated environment. He focuses on Kawauchi Village, which was once evacuated entirely due to its proximity to the Fukushima Daiichi NPP but has become the first municipality to issue the "call to return" for its residents. Decontamination of living spaces is an essential requirement for the villagers who consider returning to their homes. Thus far, houses and their premises have the highest priority of decontamination, and agricultural fields have the next priority. At present, decontamination has not been attempted in mountains and forests. The reasons seem quite obvious: mountainous areas are large and hard to decontaminate using the available techniques, and they are not considered economically productive spaces. Yet some villagers claim that mountains and forests are essential spaces for their livelihood. This chapter reveals why economically unproductive mountains and forests are perceived as socially valuable for residents of the contaminated areas.

We conclude the volume with a chapter by Yamakawa and Yamamoto (Chapter 10), which describes the cumulative impact of nuclear-disaster damage, ongoing problems, and challenges in and around the Fukushima Daiichi NPP, as well as issues that still need to be studied and addressed by policies. It is our hope that this book, along with the concurrently published volume, will allow an English-language audience to join us in reflecting on the highly complex disaster in Fukushima, in seeking ways in which the livelihoods of ordinary people can be restored, and in better preparing for future nuclear disasters, which, unfortunately, are bound to happen in some parts of the world as long as this source of power is used.

Notes

1 There have been some modifications to these designations, and on August 8, 2013, with the re-designation of the Yamakiya section of Kawamata Town, this process of tripartite division was completed.

2 Subsequently, the municipal government offices of Okuma Town were moved to Aizu-Wakamatsu City and those of Futaba Town were moved to Iwaki City (they had been in Kasu City in Saitama Prefecture until June 2013).

References

Carpenter, Susan. 2012. *Japan's Nuclear Crisis*. New York: Palgrave MacMillan.

Elliott, David. 2012. *Fukushima*. Hampshire: Macmillan.

Hindmarsh, Richard. 2013. *Nuclear Disaster at Fukushima Daiichi*. New York: Routledge.

Kingston, Jeff. 2012. *Natural Disaster and Nuclear Crisis in Japan*. New York: Routledge.

Lochbaum, David, Edwin Lyman, Susan Q Stranahan, and The Union of Concerned Scientists. 2014. *Fukushima*. New York: The New Press.

Nadesan, Majia Holmer. 2013. *Fukushima and the Privatization of Risk*. Hampshire: Palgrave Macmillan.

Samuels, Richard J. 2013. *3.11*. Ithaca, NY: Cornell University Press.

Willacy, Mark. 2013. *Fukushima*. Australia: Macmillan Publishers.

Appendix to Introduction

Kencho Kawatsu, Kenji Ohse, and Kyo Kitayama

A brief overview of basic knowledge related to radiation and radioactivity

Definitions of radionuclides, radiation, and radioactivity

"Radionuclides," "radiation," and "radioactivity" are three interrelated terms that can be best introduced and comparatively related by an analogy to a flashlight. In this analogy, radionuclides are the flashlight itself. Whereas a flashlight is a device that emits light, radionuclides (more generally called "radioactive materials"), such as radioactive iodine or radioactive cesium, are substances that emit radiation. Radiation is like the light given off from the flashlight: it refers to what is emitted by radionuclides. This emitted radiation may take the form of particle radiation—such as alpha rays (from helium nuclei) or beta rays (from electrons)—or it may take the form of electromagnetic radiation, such as gamma and X-rays. Returning once again to our analogy with a flashlight, while the "flash" in flashlight describes the ability of a flashlight to emit light, the term "radioactivity" speaks to the capacity of radionuclides to emit radiation. In regard to the strength of this capacity, whereas the power of a flashlight is measured in units of candlepower, the intensity of radioactivity is measured in units of becquerel (Bq). On the other hand, the effects or risks of radiation on the human body are represented in units of sievert (Sv), mentions of which became commonplace in the media and in various disaster-related discourses in the aftermath of the nuclear accident.

The ability of radiation to penetrate a given substance varies according to the type of radiation and the properties of the substance. For example, alpha rays can be blocked by a single piece of paper. Beta rays cannot be blocked by a piece of paper but a thin sheet of aluminum can block them. Gamma and X-rays cannot be blocked by paper or aluminum but can be blocked by a thick piece of lead. This is of course why doctors and radiologists conducting radiography-imaging inspections wear lead aprons. Lead blocks X-rays and thus provides a means of minimizing radiation exposure.

The half-life of radioactivity

Radioactive materials are structurally unstable, and they transform into structurally stable substances by emitting radiation. For example, iodine-131 emits radiation in the form of beta and gamma rays and finally transforms into the structurally stable substance known as xenon. This transformative process is called "decay." Decay is not an instantaneous transformation. For example, in the case of iodine-131, in a span of approximately eight days half of the atoms of a sample of a given quantity will have decayed into xenon, and in a span of approximately 16 days roughly 75 percent of the original iodine-131 atoms in the sample will have decayed into xenon. For this reason, and as this example begins to suggest, the length of time in which a radioactive material loses half of its radioactive atoms through decay is known as "half-life." The length of this half-life period varies according to the type of radioactive material. For example, cesium-134 has a half-life of two years, while cesium-137 has a half-life of 30 years. Both of these substances emit beta and gamma rays at the pace of their respective half-life periods and eventually transform into structurally stable barium-134 and barium-137 respectively. Therefore, cesium-131 became undetectable within a few months of the nuclear accident, and cesium-134 has been reduced to less than one fourth of the amount immediately after the accident. However, cesium-137, with its long half-life, has not been reduced as much and requires long-term countermeasures.

Natural and artificial radiation

Broadly speaking, there are two types of radioactivity and radiation: natural and artificial. We are exposed to natural radiation during the course of our normal daily lives. The worldwide average annual natural radiation exposure dose is 2.4 mSv (Table 0.1). If we break this cumulative annual natural radiation exposure dose down according to its various sources, we find that 0.39 mSv comes from outer space, 0.45 mSv from Earth, 0.29 mSv from food, and 1.26 mSv from airborne radon. Radiation exposure doses resulting from natural radiation vary from 1–10 mSv/year by country and region. For example, it is said that average annual natural radiation exposure dose in the US and Japan is 3.0 mSv and 2.1 mSv respectively.

Natural radiation is also found within the human body due to the fact that the food we consume contains small amounts of radioactive materials. For example, the body of an individual weighing 60 kg contains approximately 130 g of potassium, and around 0.0117 percent of that amount is potassium-40, which is equivalent to approximately 4000 Bq of radioactivity. Additionally, as a result of cosmic rays, a small portion of nitrogen in the atmosphere is transformed into radioactive carbon-14. This carbon-14 is brought into and subsumed into the human body through the consumption of plants and animals. This radioactivity totals about 2500 Bq for the whole body. Combined with other naturally radioactive substances, there is approximately 7000 Bq of radioactivity in the body, and

Table 0.1 Various levels of radiation and implications

Dose (mSv)	*Source/implication*
Up to 5,000	One minute's exposure to Chernobyl core shortly after explosion
8,000	Fatal dose, even with treatment
4,000	Usually fatal radiation poisoning. Survival occasionally possible with prompt treatment
2,000	Severe radiation poisoning, in some cases fatal
1,000	Causes temporary radiation sickness, including nausea and decreased white-blood-cell count
250	Upper annual limit allowed for Fukushima emergency workers
120	Average total dose received by liquidators at Chernobyl (1986–90)
100	Lowest one-year dose clearly linked to increased cancer; dose received by two Fukushima plant workers
50	Maximum yearly dose permitted by US and Japanese nuclear-industry for radiation workers
30	Average total dose of external radiation received by evacuees from Chernobyl plant and surrounding area
20	Average annual limit for nuclear-industry workers
9	One computed dose from abdomen and pelvis CT scan
9	Total exposure of airline crew flying regularly between New York and Tokyo (polar route)
4	Maximum difference of natural radiation dose in each Japanese prefecture (from Gifu to Kagawa)
3	One mammogram
2.4	Average annual background radiation globally (1.5 in Australia, 3 in North America)
Dose (μSv)	
50	Radiation dose from chest X-ray examination
40	Extra dose in Tokyo during the weeks following Fukushima accident; approximate total dose at one station at the north-west edge of the Fukushima exclusion zone
6	Dose from spending an hour on the grounds of the Chernobyl plant in 2010 (but varies widely)

Source: David W. Edgington, based on Munroe (2011).

the annual radiation exposure dose from this internal radiation is approximately 0.29 mSv.

With the exception of extraordinary circumstances of radiation exposure from atomic-weapons testing and nuclear accidents, the majority of the artificial radiation from human-induced radioactivity that people are exposed to in their daily lives comes from medical radiation exposure. For example, a single chest X-ray results in radiation exposure of 0.1 mSv, while a CT scan results in 5–30 mSv of exposure. The amount of medical radiation exposure varies widely, but is particularly high in developed countries. The average US and Japanese medical radiation exposure doses are 3.0 and 2.3 mSv/year respectively. It is essential to understand how much radiation we are exposed to in our daily lives in order to control nuclear disaster-induced radiation exposure and to assess it effects.

The effects of radiation on the human body

The effects of radiation on the human body include deterministic and stochastic effects. Deterministic effects have a threshold, a boundary value below which a certain action does not produce a certain effect and above which that action does produce that effect. Comparatively speaking, deterministic effects occur when an individual has received very high radiation exposure doses. Deterministic effects include hair loss, cataracts, and genetic disorders. In contrast, stochastic effects are assumed to be unbounded by any threshold, meaning that even very low levels of radiation exposure have the potential to bring about these effects. Examples of such stochastic effects include cancer and leukemia. According to the International Commission on Radiological Protection (ICRP), around 0.55 percent of individuals who receive 100 mSv of radiation to their whole body will later develop cancer. In Japan today, it is said that around 50 percent of people will eventually develop cancer, and it is also said that it is extremely difficult to determine whether radiation exposure or another factor is the cause of any particular instance of cancer.

Reference

Munroe, Randall. 2011. "Radiation Dose Chart." Accessed January 10, 2015. http://commons.wikimedia.org/wiki/ File:Radiation_Dose_Chart_by_Xkcd.png.

1 Shaky ground

The geophysical dynamics and sustained seismicity of the 2011 Tohoku Earthquake

Yosuke Nakamura

March 11, 2011 is the date of one of the largest earthquakes in recorded history. At 2:46 in the afternoon plate movements deep below the ocean surface in the subduction zone off the northeastern coast of Japan triggered long and violent quaking across an extremely wide area. The displacement of water caused by the initial plate movement gave rise to a tsunami that effortlessly washed over the defenses of the Tohoku coast, claiming countless victims, devastating entire communities, and resulting in one of the largest nuclear disasters the world has seen. The complex interweaving of this massive earthquake, tsunami, and nuclear accident has become known to the world as the Great East Japan Earthquake Disaster. While nearly all of the chapters in this book deal directly with this complex disaster, this chapter returns to the precipitating event—the earthquake off the Pacific coast of Tohoku (hereafter, "2011 Tohoku Earthquake")—in order to outline its geophysical sources and ongoing seismic effects.

It is not uncommon for past earthquakes to be framed as momentary, finite, and now finished events. The reality is, however, that aftershocks and other crustal movements resulting from a magnitude 9 class earthquake such as the 2011 Tohoku Earthquake can continue for well over a decade. Accordingly, it is imperative that our efforts to understand and respond to the Fukushima nuclear accident are advanced upon a foundational examination of the geophysical event that triggered this disaster and continues to sustain increased seismicity in the region. This chapter takes up this task by examining the geophysical dynamics of the 2011 Tohoku Earthquake, focusing specifically on the mechanisms of the foreshock and mainshock as well as predicted future seismicity in the form of aftershocks and induced earthquakes.

Foreshocks and mainshock

The earthquake that occurred at 2:46 pm on March 11, 2011 was an Mw9.0[1] earthquake, the fourth largest earthquake in world history since the twentieth century began. While the mainshock of this earthquake occurred off the coast of Miyagi Prefecture (Figure 1.1), around one month before this mainshock a site 50 km north of the epicenter of this mainshock became highly seismically active. Over the following weeks this seismic activity gradually shifted southward, and two

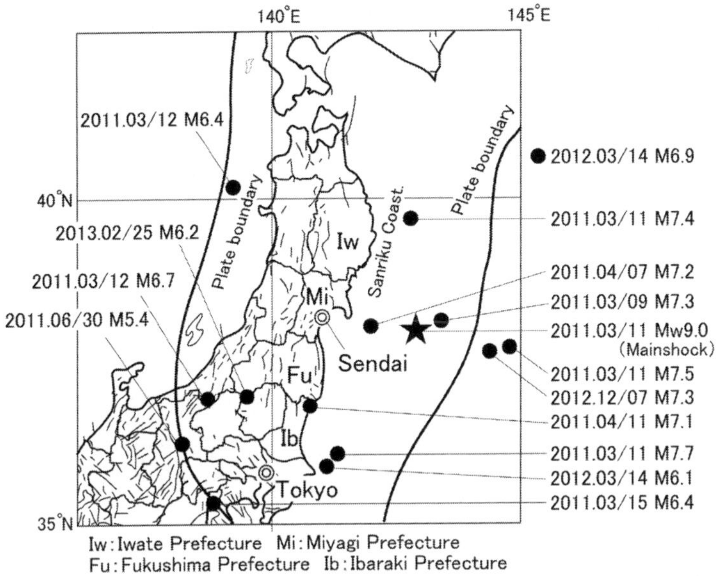

Figure 1.1 Epicenters of foreshocks, mainshock, aftershocks, and induced earthquakes related to the 2011 Tohoku Earthquake.

Source: Based on Nakamura (2014).

days before the mainshock, on March 9, 2011, a M7.3 earthquake occurred 20 km north of the epicenter of the mainshock (Kato *et al.* 2012). This earthquake produced a 50 cm tsunami that damaged oyster farms off the Sanriku Coast. The magnitude of this earthquake led seismologists to initially identify it as a mainshock. In retrospect, however, seismologists now believe that this was a foreshock of the 2011 Tohoku Earthquake.

The shaking caused by the 2011 Tohoku Earthquake was felt across almost all of Japan, excluding Okinawa Prefecture and a portion of Kyushu. On the Japan Meteorological Agency seismic-intensity scale the quake was measured as a 7—the highest level and one at which people are violently thrown by the shaking and most residences will collapse—in Osaki City in Miyagi Prefecture and as a 6-upper—the second highest level and one at which it is impossible to stand and many residences will collapse—across a wide area stretching from Iwate Prefecture to Ibaraki Prefecture. While the shaking in Tokyo was measured only as a 5-upper on the Japan Meteorological Agency seismic-intensity scale, the long duration and intensity of the shaking caused extensive damage, including bending the tip of Tokyo Tower. Frequent aftershocks followed, including a M7.5 aftershock off the coast of Iwate Prefecture 15 minutes after the mainshock and a M7.7 aftershock off the coast of Ibaraki Prefecture 30 minutes after the mainshock (Asano *et al.* 2011; Huang and Zhao 2013). Even today, over four years after the mainshock, this area continues to be highly seismically active.

Aftershocks and remotely triggered earthquakes

Following the mainshock of March 2011 a large number of small to large earthquakes occurred, primarily throughout eastern Japan. These earthquakes can be classified into several types based on their characteristics, including 1) aftershocks occurring along the plate boundary of the Japan Trench, 2) remotely triggered normal-fault earthquakes occurring inland near the boundary between Fukushima and Ibaraki prefectures, 3) remotely triggered earthquakes occurring in areas distant from the epicenter of the 2011 Tohoku Earthquake, and 4) outer-rise earthquakes occurring beyond the Japan Trench.

Earthquakes of the first type—aftershocks occurring along the plate boundary of the Japan Trench—include the M7.4 earthquake off the Iwate coast and the M7.7 earthquake off the Ibaraki coast that occurred on the same day as the 2011 Tohoku Earthquake, as well as the M7.2 earthquake that occurred off the Miyagi coast on April 7, 2011, nearly one month after the mainshock. One previous earthquake that it is highly important to consider when thinking about future seismic activity off the coast of eastern Japan is the M9.1 Sumatra Earthquake of 2004. The mainshock of this earthquake occurred on December 26, 2004, and it was followed over approximately the next ten years by over 10 earthquakes in the 7–8 magnitude class. A similar mechanism lay behind both this earthquake and the 2011 Tohoku Earthquake. Both quakes occurred at plate boundaries and were low-angle reverse-fault type earthquakes. As with the Sumatra Earthquake, frequent aftershocks as well as remotely triggered earthquakes followed the mainshock of the 2011 Tohoku Earthquake, and even now this area continues to be highly seismically active. Accordingly, it is probable that large-scale aftershocks will continue to occur for a number of years in the area off the coast of eastern Japan.

In general, the largest of aftershocks tends to be around a full digit in magnitude less than the magnitude of the mainshock; therefore, the Mw9.0 mainshock of the 2011 Tohoku Earthquake could give rise to quite large M8 aftershocks. The largest aftershocks tend to occur most frequently at the end of the fault source because the amount of fault displacement at the time of the mainshock varies according to location along the fault, with the largest amount of fault displacement tending to occur near the center of the fault. For example, the M7.0 earthquake in Sichuan, China on April 20, 2013 occurred along the same Longmenshan fault zone where the M8.0 Great Sichuan Earthquake occurred on May 12, 2008. The 2013 earthquake was likely an aftershock of the Great Sichuan Earthquake of 2008. In the case of the 2011 Tohoku Earthquake there was also a very large difference between the amount of displacement at the center of the fault (maximum 20–30 m) and the amount of displacement at the ends of the fault (less than a few meters), and thus there are concerns that a large aftershock may occur in the future at either end of the fault source off the coasts of Iwate and Ibaraki prefectures. In addition, since the southernmost aftershocks from the 2011 Tohoku Earthquake have occurred along the northern edge of the Philippine Sea Plate, it has been pointed out that this plate obstructed the southward advancement of crustal rupturing toward the coast of the Boso Peninsula. What this suggests is that, while on one hand the 2011 Tohoku Earthquake subsided with an Mw9.0 scale earthquake, on the other it has added a great amount of pressure to the northern edge of the Philippine Sea Plate.

The second type of earthquake that occurred following the 2011 Tohoku Earthquake—remotely triggered normal-fault earthquakes occurring inland—was caused primarily by tensile stress on normal faults (Figure 1.2), as represented by the high level of seismic activity that appeared in northern Ibaraki Prefecture and in the coastal Hama-Dori region of Fukushima Prefecture. While these earthquakes were nearly completely absent prior to the 2011 Tohoku Earthquake, they began to occur frequently after March 11, 2011 (Okada *et al.* 2011).

The 2011 Tohoku Earthquake was a plate-boundary earthquake, meaning that it occurred where a continental plate (i.e. the Eurasian Plate) is lifted over an oceanic plate (i.e. the Pacific Plate). As a result of this quake the fault moved roughly 50 m, and, additionally, since this is a low-angle fault, the horizontal movement of the fault was many times greater than its vertical movement (Figure 1.3). Accordingly, the areas to the west of the fault source were pulled toward the fault (the "uplift" zone in Figure 1.3). For example, the Oshika Peninsula of Miyagi Prefecture moved 5.3 m eastward as a result of this movement. However, since the absolute volume of the continental crust was not altered by this horizontal movement, the uplift of the continental crust in the vicinity of the fault induced an east–west tensile stress, "pulling in" the continental crust from the areas further west of the source fault. Consequently, the crust became thinner and subsidence resulted in these areas, which appears to be the cause of frequent normal-fault earthquakes in the northern portion of Ibaraki Prefecture and in the coastal Hama-Dori region of Fukushima Prefecture.

On April 11, 2011, one month after the 2011 Tohoku Earthquake, the M7.0 (Mw6.6) Fukushima Hama-Dori Earthquake, with its epicenter in Iwaki City, occurred, and the Hama-Dori and Naka-Dori regions of Fukushima Prefecture, as well as the southern region of Ibaraki Prefecture, experienced shaking of 6-lower

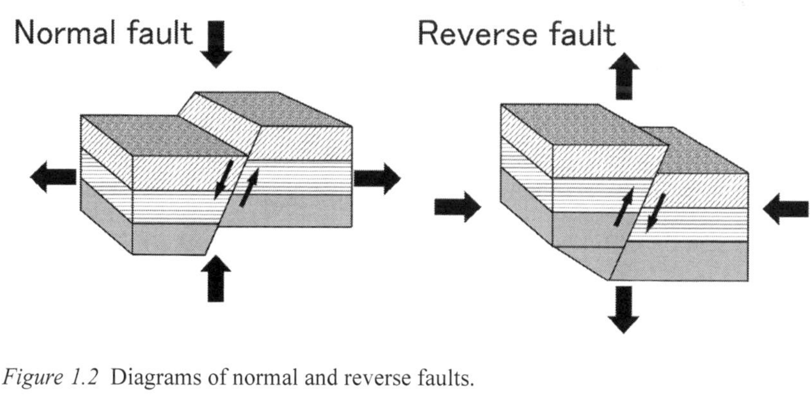

Figure 1.2 Diagrams of normal and reverse faults.

Source: Based on Nakamura (2014).

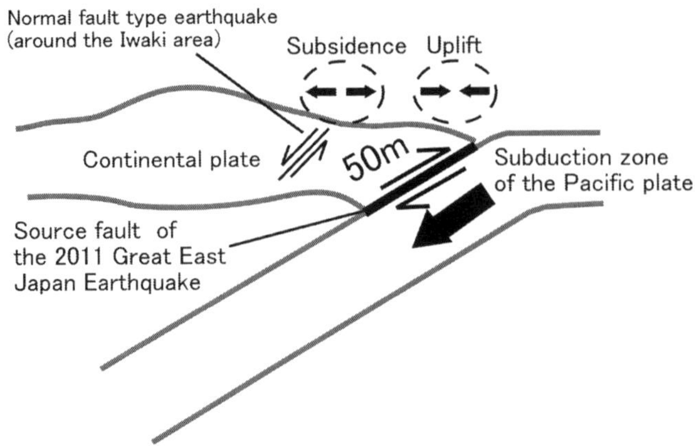

Figure 1.3 The mechanism behind the normal-fault-type earthquakes in the coastal areas of Fukushima and Ibaraki prefectures.

Source: Based on Nakamura (2014).

on the Japan Meteorological Agency seismic-intensity scale. Seismic activity remained strong in this area after the earthquake, and on September 20, 2013, roughly two and a half years after the Hama-Dori Earthquake, a M5.9 earthquake, again with its epicenter in Iwaki City, occurred, and shaking of 5-upper was recorded in Iwaki City. During the Hama-Dori Earthquake, landslides led to the deaths of four individuals, and Iwaki City experienced serious damage less than a month after the 2011 Tohoku Earthquake. This earthquake was caused by the movement of the Idosawa and Yunodake Faults, whose presence was previously only assumed from their geomorphologic characteristics. The active presence of these faults has now been confirmed by the emergence of a clear normal fault along west of the Idosawa Fault (Lin *et al.* 2013). The results of a trench-excavation survey of the normal fault created by this earthquake indicated that the most recent activity of this fault prior to the Hama-Dori earthquake was 12,620 to 17,410 years ago (Toda and Tsutsumi 2013). Because it was not possible to find traces of movement at the time of the Jogan Earthquake of 869, it can be said that no earthquakes have occurred on this fault as a result of large earthquakes occurring at the Japan Trench. This indicates that the faults along the Idosawa Fault have been moving independently of large earthquakes along the Japan Trench, suggesting that due to conditions at the time a fault may have been moving parallel to the Idosawa Fault.

The third type of earthquake that occurred following the 2011 Tohoku Earthquake—remotely triggered earthquakes in areas distant from the epicenter of the mainshock—includes such earthquakes as the M6.7 earthquake that occurred the following day in northern Nagano Prefecture, the M6.4 earthquake that occurred on March 15, 2011 in eastern Shizuoka Prefecture, the M5.4 earthquake that occurred on June 30, 2011 in central Nagano Prefecture, and the

M6.2 earthquake that occurred on February 25, 2012 in northern Tochigi Prefecture. During the northern Nagano Earthquake of March 12, 2011, Sakae Village, located near the epicenter of the quake, experienced shaking of 6-upper on the Japan Meteorological Agency seismic-intensity scale, and the Tsunan area of Tokamachi City experienced shaking of 6-lower. While this earthquake forced 1,700 individuals from Sakae Village, over 80 percent of the population, to evacuate, the high degree to which public attention across the nation was centered on the 2011 Tohoku Earthquake Disaster meant that the impacts of this disaster went largely unreported by the media. The eastern Shizuoka Earthquake that occurred four days after the 2011 Tohoku Earthquake was recorded as a 6-upper shaking in Fujinomiya City, Shizuoka Prefecture, and 50 individuals were injured. Since the epicenter of this earthquake was directly over the magma chamber of Mt Fuji, volcanologists were concerned that an eruption was immanent. Fortunately, however, no eruption of Mt Fuji has yet occurred following the 2011 Tohoku Earthquake.

The central Nagano Earthquake of June 30, 2011 occurred directly under the highly populated city of Matsumoto. Over 4,000 cases of structural damage were reported, and one individual lost their life. While this earthquake was initially thought to have been generated by the active Gofukuji fault of the Itoigawa Shizuoka Tectonic Line, detailed analysis of the location of the epicenter indicated that it occurred at the far west of the Akagi Fault, which branches off from the Gofukuji Fault (it is possible that these two faults are connected underground). In the case of the earthquake that struck northern Tochigi Prefecture on February 25, 2012, shaking of 5-upper on the Japan Meteorological Agency seismic-intensity scale was recorded, and an avalanche resulting from the earthquake caused guests at a hot-springs resort to be temporarily blocked in. Mt Nasu and Mt Nikko-Shirane are located in this area, and it is one of the areas in Japan with the highest number of volcanoes. Since the 2011 Tohoku Earthquake resulted in increased seismic activity in an area stretching from Hokkaido to Kyushu that contains 20 volcanoes (in 2014, two of these, Mt Ontake and Mt Aso, erupted), it is imperative to continue to observe this seismic activity in volcanic regions.

Outer-rise earthquakes

The fourth category of earthquake that followed the 2011 Tohoku Earthquake—outer-rise earthquakes occurring beyond the Japan Trench—include the M7.5 earthquake that occurred approximately 40 minutes after the mainshock, 100km beyond the Japan Trench, as well as the M7.3 earthquake that occurred in the same area on December 7, 2012 (Harada *et al.* 2013).

The "rises" that this term "outer-rise earthquake" references are formed in the area where oceanic plates begin to slide under continental plates (i.e. the outer-rises), and the shallow portions of the plate stretch as the plate bends in a downward direction (Figure 1.4). Accordingly, an outer-rise earthquake generally has a shallow epicenter and is a normal-fault earthquake (Sleep 2012). Additionally, as a result of the influence of the destruction of a fault caused by a reverse-fault plate-boundary earthquake, outer-rise earthquakes often occur in tandem with a mainshock. While

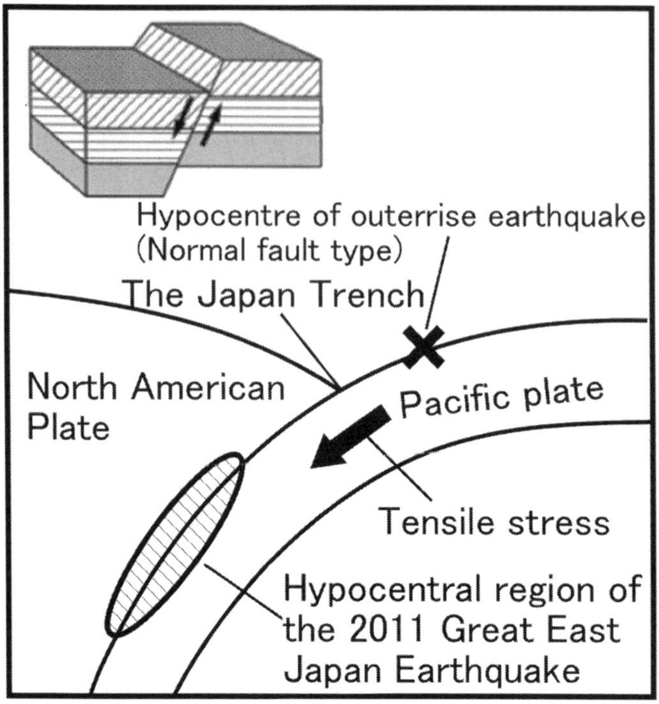

Figure 1.4 Diagram of an outer-rise earthquake.

Source: Based on Nakamura (2014).

an outer-rise earthquake could potentially occur soon after a mainshock, at other times an outer-rise earthquake will occur a full decade after the mainshock. In general, because these earthquakes occur far from inland areas, the shaking from these earthquakes is limited, yet their defining feature is that they often trigger large tsunami. Furthermore, outer-rise earthquakes can occur many years after a mainshock and can be of approximately the same magnitude as the mainshock. Since it is not possible to say that all of the force added by the 2011 Tohoku Earthquake has been released, it is imperative to be cautious of an outer-rise earthquake occurring in the future.

One important example of an outer-rise earthquake is the Showa Sanriku Earthquake of 1933. This M8.1 normal-fault outer-rise earthquake was caused at least partially by the M8.2 Meiji Sanriku Earthquake of 37 years earlier. The Meiji Sanriku Earthquake of June 15, 1896 was a reverse-fault plate-boundary earthquake of M8.2 with an epicenter 200 km off the coast of Kamaishi City, Iwate Prefecture. Since the epicenter of the earthquake was far from the mainland, the damage caused by the quake was relatively light. However, the tsunami resulting from the earthquake was measured at 38.2 m, the largest to hit the island of Honshu since measurements began. Over 20,000 individuals were killed or went missing as a result of this tsunami, and nearly 10,000 homes were washed away. Since this was an earthquake of very large scale, remotely triggered earthquakes later occurred far from the epicenter of the quake and, two and a half months after the Meiji Sanriku Earthquake on August 31, 1896, the M7.2 Rikuu Earthquake occurred along the Senya Fault of Akita Prefecture, and large amounts of damage were experienced across the Yokote Basin. Additionally, eight months later, in February 1897, an M7.4 earthquake occurred off the coast of Miyagi Prefecture. Following this earthquake, 43 villages of Iwate and Miyagi Prefecture were relocated to higher ground. This process of relocating communities to higher ground has occurred several times in the past following tsunamis along the coasts of Iwate and Miyagi prefectures, and the 2011 Tohoku Earthquake has similarly led to relocation efforts, this time on an unprecedented scale.

The Showa Sanriku Earthquake occurred on March 3, 1933 and was an M8.1 earthquake with an epicenter 200 km off the coast of Kamaishi City in Miyagi Prefecture. Importantly, the epicenter of this quake was located further east than that of the Meiji Sanriku Earthquake. As in the case of the Meiji Sanriku Earthquake, the damage from the earthquake was relatively minor. Yet the energy from the earthquake was very large and it triggered an enormous tsunami that resulted in major damage to the area. The largest waves of 28.7 m were recorded in Ryori Village, Kesen District (present-day Ofunato City). Over 3,000 individuals were killed or went missing, and nearly 5,000 homes were destroyed. Since this was a normal-fault earthquake occurring in the outer edges of the trench uplift zone and because its epicenter was much further east than the Meiji Sanriku Earthquake it can be argued that this was an outer-rise earthquake.

Future earthquake predictions

Magnitude 9 class earthquakes around the world have in the past resulted in very large aftershocks many years after the mainshock. For example, if we look at the case of the M9.1 Sumatra Earthquake of 2004 we see that magnitude 7–8 class aftershocks continue to occur annually even today, over ten years after the mainshock. Accordingly, it can be expected that aftershocks and remotely triggered earthquakes from the 2011 Tohoku Earthquake, which closely resembled the Sumatra Earthquake, can be expected to continue for at least a decade after the mainshock. Additionally, the M8.4 Jogan Earthquake off the Sanriku Coast in 869, the last massive earthquake in the same fault zone that caused the 2011 Tohoku Earthquake, resulted, nine years later, in the M7 class epicentral Sagami-Musashi Earthquake and, 18 years later, in the M8 class Niwa Earthquake along the Nankai Trough. While the Heian Era was a period of great continental-crust seismic activity, when earthquakes occurred all over Japan, it is possible that, similar to the Jogan Earthquake of the Heian Era, that the occurrence of the 2011 Tohoku Earthquake, and the increased seismicity it has brought along with it, may again result in an epicentral earthquake beneath Tokyo or a remotely triggered earthquake along the Nankai Trough, both of which have been estimated as leading to numerous casualties and economic losses of over a trillion dollars. In addition, large plate-boundary earthquakes have in the past resulted in inland fault earthquakes as well as volcanic eruptions. In 2014, both the M6.7 Kamishiro Earthquake and eruptions at Mt Ontake and Mt Aso occured. It is very likely that remotely triggered earthquakes and volcanic eruptions will continue over the next few years. Accordingly, there is a serious need to study and learn from past earthquakes and volcanic activities.

In summary, the 2011 Tohoku Earthquake was a massive Mw9.0 earthquake, and aftershocks and remotely triggered earthquakes can be expected to continue for over a decade. To understand the geophysical dynamics and sustained seismicity, it is critical to learn from past geological events, which help us to remember that an earthquake can occur at any time and continuous preparation is critical. Indeed, there is ample potential for an epicentral earthquake below the Tokyo metropolitan area and an earthquake along the Nankai Trough. The occurrence of these earthquakes could potentially impact the process of resolving the Fukushima nuclear fallout and the operation of other nuclear plants in the country. Accordingly, it is imperative to remember the potential for large-scale earthquakes to occur while conducting recovery efforts.

Note

1 Regarding the usage of "magnitude" in this chapter, "Mw" will be used to refer to "moment magnitude," while "M" is used to refer to the Japanese Metrological Agency's "magnitude" scale.

Acknowledgment

This work was supported by JSPS KAKENHI Grant Number 25220403.

References

Asano, Yoichi, Tatsuhiko Saito, Yoshihiro Ito, Katsuhiko Shiomi, Hitoshi Hirose, Takumi Matsumoto, Shin Aoi, Sadaki Hori, and Shoji Sekiguchi. 2011. "Spatial Distribution and Focal Mechanisms of Aftershocks of the 2011 off the Pacific Coast of Tohoku Earthquake." *Earth Planets Space*, 63(7): 669–73.

Harada, Tomoya, Murotani Satoko, and Kenji Satake. 2013. "A Deep Outer-rise Reverse-fault Earthquake Immediately Triggered a Shallow Normal-fault Earthquake: the 7 December 2012 off-Sanriku Earthquake (Mw 7.3)." *Geophysical Research Letters*, 40(16): 4214–19, doi 10.1002/grl.50808.

Huang, Zhouchuan, and Dapeng Zhao. 2013. "Relocating the 2011 Tohoku-oki Earthquakes (M 6.0–9.0)." *Tectonophysics*, 586: 35–45.

Kato, Aitaro, Kazushige Obara, Toshihiro Igarashi, Hiroshi Tsuruoka, Shigeki Nakagawa, and Naoshi Hirata. 2012. "Propagation of Slow Slip Leading Up to the 2011 Mw 9.0 Tohoku-Oki Earthquake." *Science*, 335(6069): 705–8, doi:10.1126/science.1215141.

Lin, Aiming, Shinji Toda, Gang Rao, Satoru Tsuchihashi, and Bing Yan. 2013. "Structural Analysis of Coseismic Normal Fault Zones of the 2011 Mw 6.6 Fukushima Earthquake, Northeast Japan." *Bulletin of the Seismological Society of America*, 103(2B): 1603–13, doi:10.1785/0120120111.

Nakamura, Yosuke. 2014. "Outline of the 2011 off the Pacific Coast of Tohoku Earthquake and Future Prediction of Earthquake Occurrence in Japan." In *Higashinihon Daishinsai kara no Fukkyu/fukkou to Kokusaihikaku* [Reconstruction from 2011 Tohoku Earthquake and Tsunami], edited by the Fukushima University International Disaster Reconstruction Research Group, 143–58. Tokyo: Hassaku-sha. [In Japanese.]

Okada, Tomomi, Keisuke Yoshida, Sadato Ueki, Junichi Nakajima, Naoki Uchida, Toru Matsuzawa, Norihito Umino, Akira Hasegawa, and Group for the Aftershock Observations of the 2011 Off the Pacific Coast of Tohoku Earthquake. 2011. "Shallow Inland Earthquakes in NE Japan Possibly Triggered by the 2011 off the Pacific Coast of Tohoku Earthquake." *Earth Planets Space*, 63(7): 749–54, doi:10.5047/eps.2011.06.027.

Sleep, Norman H. 2012. "Constraint on the Recurrence of Great Outer-rise Earthquakes from Seafloor Bathymetry." *Earth Planets Space*, 64(12): 1245–6.

Toda, Shinji, and Tsutsumi, Hiroyuki. 2013. "Simultaneous Reactivation of Two Subparallel Inland Normal Faults during the Mw 6.6 11 April 2011 Iwaki Earthquake Triggered by the Mw 9.0 Tohoku-oki, Japan, Earthquake." *Bulletin of the Seismological Society of America*, 103(2B): 1584–1602, doi:10.1785/0120120281.

2 Outline of an invisible disaster

Physio-spatial processes and the diffusion and deposition of radioactive materials from the Fukushima nuclear accident

Kencho Kawatsu, Kenji Ohse, and Kyo Kitayama

Introduction

Radionuclides released as a result of the accident at the Fukushima Daiichi Nuclear Power Plant (NPP) were deposited across a wide area of Fukushima Prefecture, resulting not only in concerns about the health effects for residents from radiation exposure but a string of interrelated and equally complex issues, including contamination and stigmatization of the agricultural and marine products of the region, restrictions on residence and entry in areas around the plant, and the division of families and communities over the difficulties of evacuation and compensation. In comparison with other prefectures severely affected by the Great East Japan Earthquake Disaster, recovery in Fukushima Prefecture has been delayed due to the major obstacles presented by the NPP accident—the invisible but highly destructive, complex, and challenging consequences of radiation contamination. In order to understand the socio-economic and cultural consequences of radiation contamination, and to gain insight into the prospects for recovery, it is imperative to first solidify our understanding of the "invisible" processes at work, as well as the extent and magnitude of radioactive contamination on the ground. Accordingly, the present chapter aims to provide a basic overview of the physical properties and dynamics of radionuclides and radiation, and the types and amount of radioactive substances released as a result of the accident. It also discusses the atmospheric transport and deposition of radionuclides and their terrestrial movement and transfer to crops, the effects of radiation on the human body, and countermeasures to respond to the NPP accident such as safety limits for radiation in food.

The situation and emitted radionuclides of the Fukushima Daiichi NPP accident

A timeline of the nuclear accident

At 2:46 pm on March 11, 2011, an earthquake occurred off the coast of the Tohoku region of northern Japan followed by an extremely large tsunami that struck the Fukushima Daiichi NPP and resulted in complete loss of power and the loss of the ability to cool the nuclear reactors at the plant. Temperatures and pressure inside the nuclear reactors subsequently surged, and venting operations and hydrogen generation resulted in the explosion of the reactor building and the release of large amounts of radioactive material. The timeline of the main events is shown in Table 2.1.

Figure 2.1 presents changes in the observed air radiation dose rate at locations throughout Fukushima Prefecture during the above sequence of events. The sharp peaks depicted in the figure indicate the rapid increases in observed air radiation dose rate as radioactive plumes (i.e. clouds containing radioactive materials) passed over the area. When materials from the radioactive plume were deposited onto the ground by snow and rain, these deposited radioactive materials began to

Table 2.1 Main events of the Fukushima Daiichi NPP accident

March 11	15:42	Complete power loss at the nuclear power plant
March 12	10:17	Venting operations at Reactor 1
	15:36	Explosion at Reactor 1
March 13	08:41	Venting operations at Reactor 3
	11:00	Venting operations at Reactor 2
March 14	11:01	Explosion at Reactor 3
March 15	06:10	Explosions at Reactors 2 and 4
	09:38	Fire at Reactor 4
March 21	18:22	Water vapor at Reactor 2

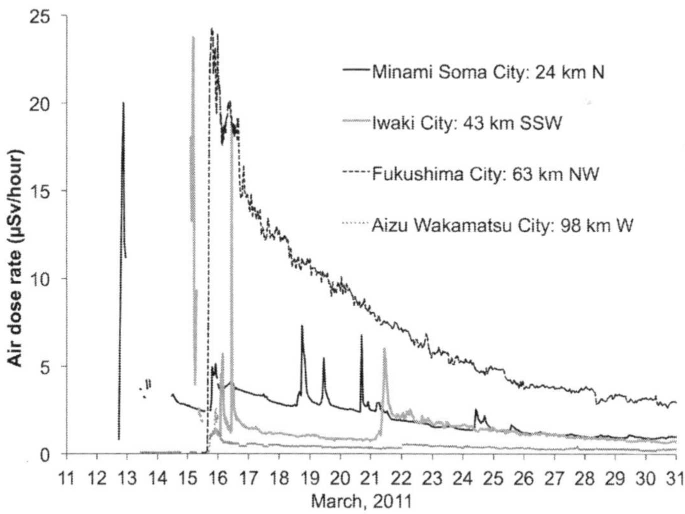

Figure 2.1 Air dose rate observed at several sites in Fukushima.

Source: Fukushima Prefecture (2014).

gradually decay at rates consistent with their half-life. Since iodine-131, which has a relatively short half-life of eight days, was heavily deposited after the accident, the initial dose rate quickly decreased (see the Appendix for the basic properties of radionuclides and radiation). Later, however, after the large deposits of radioactive iodine had become stable non-radioactive iodine, the remaining radioactive materials consisted predominantly of cesium-134 and cesium-137. Since these radioactive materials have much longer half-life periods, the initially quick decrease in the air radiation dose rate leveled off as these materials began their long process of decay.

The emissions of radionuclides from Fukushima Daiichi

As a result of the earthquake- and tsunami-induced accident at the Fukushima Daiichi NPP large volumes of radioactive materials were released into the atmosphere. Among these, the most prominent were noble gas xenon-133, iodine-131, cesium-134, and cesium-137. Table 2.2 provides a list of estimated amounts of major radioactive materials released as a result of the nuclear accident. Comparing the emissions of radioactive materials following the Fukushima accident with the emissions from the Chernobyl accident, we find that twice as much xenon-133 was emitted in the case of the Fukushima accident, the quantities of iodine-131 released were about 10 percent of Chernobyl, and the quantities of cesium-134 and cesium-137 about 20 to 40 percent (Table 2.2).

While many types of radionuclides are generated through physical phenomenon, such as nuclear fission and neutron capture in nuclear testing and within nuclear reactors, in the case of the Fukushima nuclear accident the main issues have been with radioactive iodine and cesium. In the case of the Chernobyl accident, strontium was also dispersed in large volumes and was a major problem.

Table 2.2 Comparison of radionuclide emissions from the Fukushima and Chernobyl Nuclear Power Plants

	Amount of emission (Bq)		
	Fukushima		*Chernobyl*
Radionuclides	*Nuclear Safety Commission (2011)*	*Stohl et al. (2012)*	*IAEA (2006)*
^{133}Xe	1.1×10^{19}	1.5×10^{19}	6.5×10^{18}
^{131}I	1.6×10^{17}		1.8×10^{18}
^{132}I	1.3×10^{13}		
^{133}I	4.2×10^{16}		9.1×10^{17}
^{134}Cs	1.8×10^{16}		4.7×10^{16}
^{137}Cs	1.5×10^{16}	3.7×10^{16}	8.5×10^{16}

One of the unique characteristics of these radionuclides is that their boiling point is relatively low and they are therefore relatively easily released during an accident. The boiling point of iodine is 184°C and the boiling point of cesium is 671°C, relatively low. Accordingly, these materials were emitted in great volume into the atmosphere as a result of the Fukushima nuclear accident. In contrast, the boiling point of strontium is 1382°C, and in the case of the Chernobyl accident, where the reactor itself caught fire, this substance was emitted in large volumes and contaminated a wide area. However, in the case of the Fukushima accident, where fire in the reactor structure and hydrogen explosions were the cause, temperatures were not that high, and it can be thought that strontium emissions were low. Additionally, since the half-life of these nuclides extends from a few days to decades, they can affect the health of plants and animals during the course of their life span. Most of the radioactive nuclides generated within nuclear reactors have very short half-lives. Indeed many have half-lives of only a few seconds to a few minutes. Even if these nuclides are released, they will quickly decay and their effects will be limited. On the other hand, while some nuclides have extremely long half-lives, because they do not emit large amounts of radiation into the environment they are less of a problem. Finally, animals and plants absorb and incorporate some nuclides more easily than others. A large emission of iodine-131, cesium-134, and cesium-137 is highly problematic because their half-lives are long (i.e. at least a few days), they emit large amounts of radiation, and they are easily absorbed by living organisms.

According to the results of airborne monitoring conducted by the Ministry of Education, Culture, Sports, Science and Technology (MEXT), the cesium-134 and cesium-137 emitted from the NPP were heavily deposited to the northwest of Fukushima Daiichi, from the Naka-Dori region of central Fukushima Prefecture south to the Kanto region, especially Ibaraki, Tochigi, Gunma, and Chiba Prefectures (Figure 2.2). In total, airborne monitoring had been conducted seven times by November 2013. Comparing the results of the fourth airborne monitoring of November 2011 with the results of the seventh monitoring indicates a 47 percent decrease in the air radiation dose rate, a greater reduction than was expected according to estimates of a 34 percent reduction based on physical half-life. Environmental radiation monitoring and mesh surveys of the residential areas (e.g. along the streets) of Fukushima Prefecture conducted by citizens indicated a reduction in the average air radiation dose rate of 40 percent between the first monitoring of April 2011 and the fifth monitoring of October 2012. However, since the contribution of cesium-134, which has a relatively short half-life, to air radiation dose rates will decrease in the future, the contribution of cesium-137 to the air radiation dose rate will become proportionally greater, resulting in a decrease in the rate of reduction.

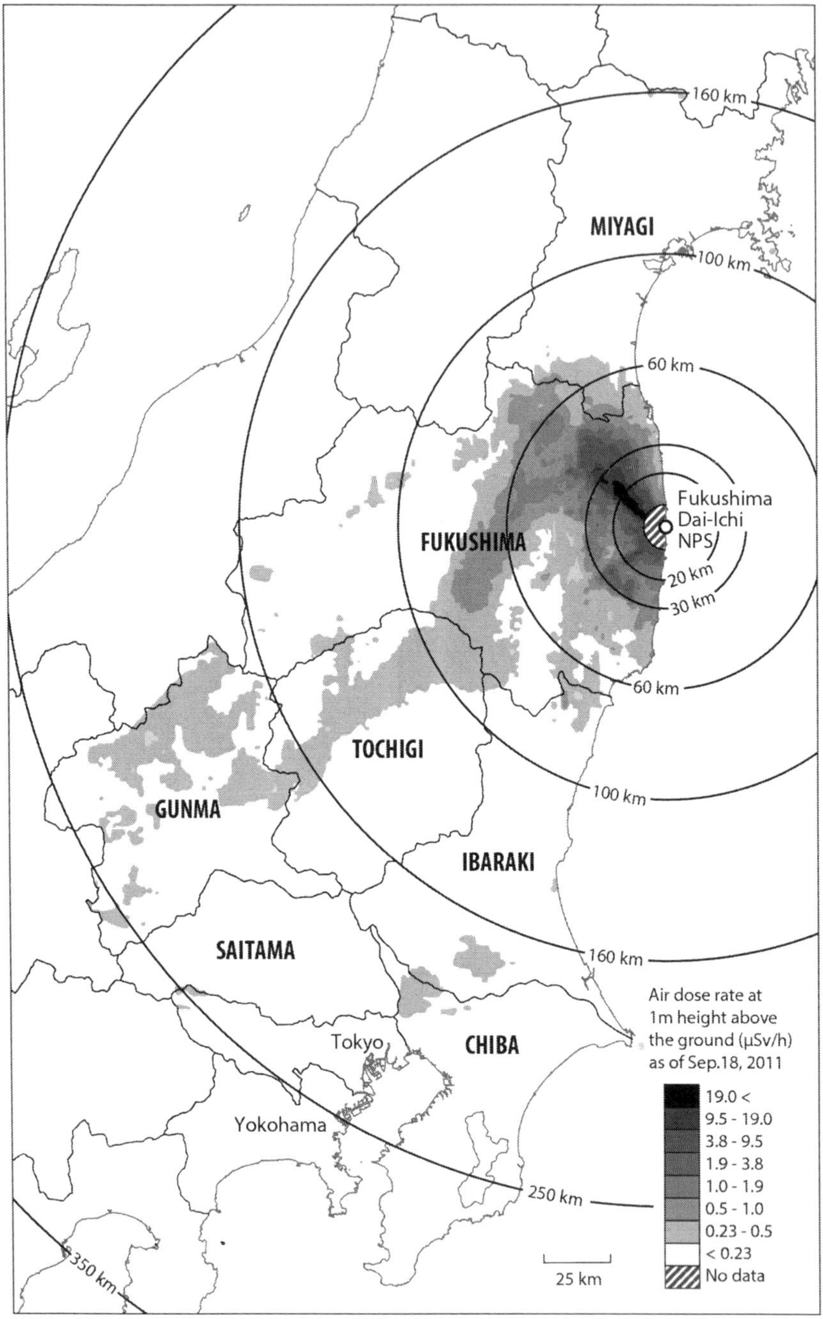

Figure 2.2 Radioactive pollution caused by the accident at TEPCO's Fukushima Daiichi NPP.

Source: Created by Eric Leinberger, based on Ministry of Environment (2013).

The atmospheric transport and dispersion of radioactive materials

The dispersion of radioactive materials through atmospheric transport and deposition

The dispersion of radioactive materials released as a result of the nuclear accident is the primary way of determining the extent and distribution of the radioactive contaminated area. The process can be divided into two main processes: atmospheric transport and deposition, and terrestrial and aquatic movement. Here we focus on atmospheric transport and deposition. In general, atmospheric transport and deposition are highly dependent on the form of a substance. Among the main radioactive materials emitted, idoine-131 exists as a gas and as particulate matter in the atmosphere, while cesium-134 and cesium-134 exist as particulate matter. As noted above, xenon-133 exists as a gas, but its reactivity is poor, meaning that it does not accumulate in the environment. Atmospheric transport is propelled by atmospheric airflows and both direction and range are determined by advection. When a lot of radionuclides were released, prevailing winds were from the northwest in the area around the Fukushima Daiichi NPP (Japan Meteorological Agency 2015). Accordingly, released radioactive materials were carried toward the Pacific Ocean, and the dispersion within the Japanese mainland has been limited.

There are two types of process through which radionuclides in the atmosphere are deposited: wet and dry deposition. Wet deposition is a process mediated by rain and snow; therefore, the amount of the deposition is influenced the amount and frequency of rainfall. In contrast, dry deposition is not mediated by rain or snow, but by molecular diffusion, turbulent dispersion, and reactions with materials on the ground that lead to deposition. In the case of the Fukushima accident, wet deposition was the major process for the contamination of the ground.

The global patterns of dispersion

Existing modeling and simulation studies that examine the atmospheric dispersion of radioactive materials indicate that radioactive materials emitted from Fukushima on March 11 reached North America within approximately one week, Europe a few days later, and by April were dispersed throughout the Northern Hemisphere (Lujaniene *et al.* 2012; Stohl *et al.* 2012; Winiarek *et al.* 2012; Evangeliou *et al.* 2013). While there are differences among different types of radionuclides in terms of the reduction of their atmospheric concentrations due to different rates of deposition, there are no major differences in their arrival times.

For terrestrial environments outside of Japan, due to decreased concentrations as a result of dispersion, the effects of the transport and deposition of radionuclides are small. The trends of the dispersion appeared in the observed data in the previous reports (Table 2.3). The peak values of the radionuclide concentrations decreased with eastward distance from the Fukushima Daiichi NPP. In particular, concentrations of radioactive materials in the atmosphere decreased drastically

Table 2.3 Atmospheric concentration peaks of particulate iodine-131 and cesium-137 in global sites immediately after the Fukushima Daiichi NPP accident

		Particulate ^{131}I		*^{137}Cs*	
Sources	*Sampling site*	*Date*	*Level (mBq/m^3)*	*Date*	*Level (mBq/m^3)*
Chino *et al.* (2011)	Around the FDNPP	March 15–21	800–2800	March 15–21	150–350
US EPA (2012)	Oahu, US (21.46°N, 158.01°W)	March 21	24	March 21	4.4
	Boise, US (43.63°N, 116.21°W)	March 23	31	March 23	2.3
	Nome, US (64.50°N, 165.41°W)	March 24	16	March 24	4.8
Lozano *et al.* (2011)	Huelva, Spain (37.27°N, 6.94°W)	March 28–29	3.7	March 28–29	0.95
Masson *et al.* (2011)	Central Europe	March 28–April 4	1–6	March 28–April 4	0.75–1
Hsu *et al.* (2012), Huh *et al.* (2012), Huh *et al.* (2013), Long *et al.* (2012)	Hong Kong, Philippines, Taiwan, Vietnam	April 4–14	0.1–1	April 4–14	0.5–1.5
Hong *et al.* (2012), Kim *et al.* (2012)	Korea	April 6	0.5–3.5	April 6	1–1.5

when they crossed the Pacific Ocean. Some westward transport was confirmed in East Asia (for example, Hong Kong and South Korea) in April, but the peak values were low because of the low emission in the period.

The local patterns of dispersion within Japan

When we focus on the patterns of atmospheric dispersion within the land areas of Japan, iodine-131's trends were similar to cesium-134 and cesium-137, flowing inland on winds flowing to the northwest and southwest. However, since iodine-131 was in a gaseous state, the greater percentage was deposited through dry deposition rather than wet deposition, and its deposition trends differ from cesium-134 and cesium-137, being deposited more heavily in a southwest direction (Ohara *et al.* 2011; Morino *et al.* 2011; Katata *et al.* 2012b). Although iodine-131 was emitted in larger volumes than radioactive cesium, it is the long-term effects of contamination by radioactive cesium that are of greater concern because of its longer half-life. Results from SPEEDI (the System for Prediction of Environmental Emergency Dose Information), operated by MEXT and WSPEEDI-II (the world wide version of SPEEDI), which was developed by the Japan Atomic Energy Research and Development Organization (JAEA), predicted that on the morning of March 15, 2011 cesium-137 would be transported by winds in a southwest direction from the NPP and primarily dry deposited, and in the afternoon it would be transported in a northwest direction and wet deposited by precipitation as far as central Fukushima and in a wide area (Chino *et al.* 2011; Terada *et al.* 2012; Katata *et al.* 2012a; Nagai *et al.* 2014). In the case of other events, dry deposition occurred in the vicinity of the NPP, while wet deposition occurred across a far wider area. Additionally, based on modeling, Terada *et al.* (2012) and Katata *et al.* (2014) suggest that deposition spread across Tochigi and Gunma Prefectures as a result of deposition by fog and mist. Other modeling results also indicate the deposition of cesium-137 across a wide range northwest of the NPP and in a smaller area to the southwest in the vicinity of the Fukushima Daiichi NPP (Morino *et al.* 2013; Draxler *et al.* 2015).

Significant damage notwithstanding, one may say that Japan and neighboring countries were lucky to avoid even greater radioactive contamination. If we compare the deposition amounts of cesium-137 after the Chernobyl and Fukushima accidents, the former resulted in a smaller amount of deposited cesium-137 than in the highly contaminated areas of Fukushima, but the area of contamination extended east and west from the nuclear accident and across the European region (De Cort 1998). Differences in the range of contamination are certainly a result of differences in wind direction in relation to the location of the accidents. The Fukushima accident occurred at a time when prevailing winds were from the northwest and west (i.e. in the direction of the Pacific Ocean). Despite occasional winds from the east that blew inland, in comparison with Chernobyl, the highly contaminated area was limited to the vicinity of the NPP. Moreover, since the prevailing winds were toward the Pacific Ocean, the contamination of other countries was slight and the majority of contamination occurred in the Pacific Ocean. Short-term precipitation resulted in deposition on land and contamination, but climatic conditions after the Fukushima accident limited the ground area of contamination.

Movement of radioactive materials in terrestrial environments and their effects on agriculture

Behavior of radionuclides observed after the Fukushima Daiichi accident

While radioactive nuclides released as a result of the Fukushima accident were transported by the atmosphere and deposited on soil and crops, the specific dynamics varied according to the chemical properties of each nuclide. Here, we will discuss short-term observations of these dynamics in Tsukuba City, approximately 170 km south of the Fukushima Daiichi NPP.

Figure 2.3 lists observed concentrations of radioactive iodine and cesium in precipitation, harvested spinach from experiment fields, and soil surface in Tsukuba City from March to July 2011. In Tsukuba City, the radioactive plume passed overhead on March 15 and the observed air dose rate was the highest on that day (Sanami *et al.* 2011), although no precipitation was recorded. Prior to that day no concentrations of radioactive iodine had been found in spinach, but spinach harvested on the morning of March 15 was found to contain over 10,000 Bq/kg. Spinach harvested on the following day was found to contain 12,900 Bq/kg, the highest value recorded during observations. These results indicate that airborne radioactive iodine was deposited directly onto spinach crops. The first rain after the accident came on March 21, and this rainwater contained over 3000 Bq/L of radioactive iodine and approximately 400 Bq/L of radioactive cesium. While the concentration of radioactive cesium decreased below 100 Bq/kg from March 15 until the rain on March 21, the rain caused the level of concentration to again rise to 1000 Bq/kg. This area of southern Ibaraki Prefecture, around Tsukuba City, and extending into northeastern Chiba Prefecture is what has been called a "hotspot," an area that while far from the reactor has high air dose rates. The cause of this hotspot was that rain fell while atmospheric concentrations of radioactivity were

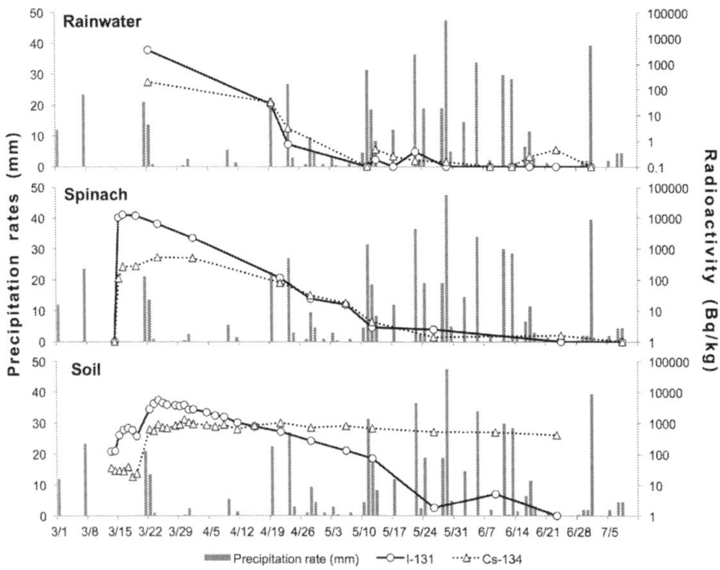

Figure 2.3 Changes in concentration of iodine-131, cesium-134, and cesium-137 in: (a) precipitation, (b) spinach, and (c) soil surface (0–5 cm) collected from upland, observed in Tsukuba City. The concentrations of radionuclides were corrected to the sampling date.

Source: Ohse *et al.* (2015).

high, resulting in the localized deposition of large amounts of radioactive cesium (Ohse *et al.* 2015).

The dynamics of radioactive cesium in terrestrial environments

Radioactive cesium had been deposited on ground surfaces even before the Fukushima Daiichi accident as a result of such events as atmospheric nuclear testing, and, based on these previous experiences, it was known that the movement of cesium within agricultural and forest ecological systems is influenced by types and patterns of land use. Figure 2.4 presents results obtained through inspection of the vertical distribution of cesium-137 in the soil of forests, dry fields, and paddy fields in the vicinity of the NPP. In forests where the ground is covered with leaves and artificial disturbances are minimal, most cesium-137 remains on the top layer and is rarely detected deeper in the soil. Additionally, outflow due to soil and wind erosion is also low. On agricultural land (dry and paddy fields), as a result of the cultivation of the soil, cesium-137 can be detected deeper in the soil than in forests. The reason for this is that while cultivation normally only happens down to 10–15 cm, farmers sometimes plow the soil even deeper. It may be that when this happens cesium-137 moves deeper into the soil. Additionally, in agricultural fields where the surface is not covered by any protective layer, erosion reduces the concentration of radioactive materials in the surface layer. This tendency is particularly prominent in rice paddies because flooding and tilling of the soil before planting sods results in fine-grained cesium particles being removed by flowing water. For this reason the overall amount of cesium-137 is less than in other land uses (Figure 2.4). In sum, for the same size area, the amount of cesium-137 is greatest in forests, followed by dry fields and paddy fields (Ohse *et al.* 2012).

Transfer of radioactive cesium to agricultural crops and countermeasures to prevent it

In 2011, radioactive cesium deposited from the atmosphere onto leaves and bark entered, through a process known as "commutation," into the edible portions of plants and animals, and relatively high concentrations were the result. At present, airborne radioactive cesium is extremely low and effects from direct absorption from the atmosphere are negligible. However, in parts of some fruit trees

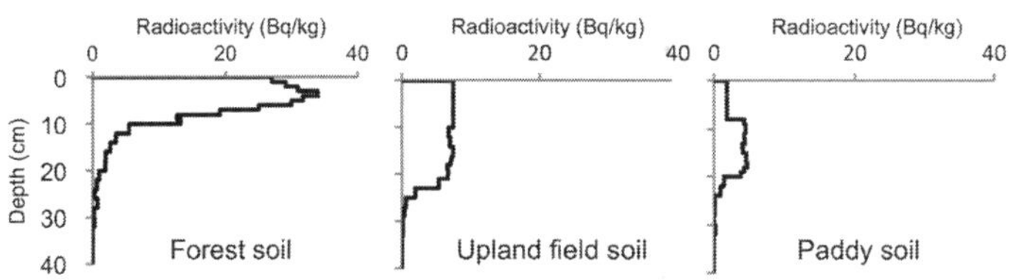

Figure 2.4 Vertical profiles of cesium-137 in forest, upland, and paddy soils.

Source: Ohse *et al.* (2012).

materials absorbed after the accident remain inside the trees, and contaminated fruits have been discovered.

Radioactive cesium deposited onto soil is transferred to crops by root absorption. In this process of transfer from soil to crops, the presence of potassium, which is an essential element for plants and is also, like cesium, an alkali metal, exerts a significant influence. If exchangeable potassium, a substance easily absorbed by plants, is present in sufficient quantities in soil then the transfer of radioactive cesium to crops will become more difficult. Additionally, radioactive cesium that is absorbed will not be evenly distributed throughout a plant: concentrations in different parts of the plant will differ. While there is a tendency in many plants for concentration in leaves to be relatively high and for concentration in seeds to be relatively low, these ratios differ according to the type of plant. The following is the transfer coefficient (*TF*), a formula used to indicate the degree of transfer of radioactive cesium (^{137}Cs) to the edible portions of plants from soil:

$$TF = \frac{\text{Concentration of } ^{137}\text{Cs in crops (Bq/kg)}}{\text{Concentration of } ^{137}\text{Cs in soil (Bq/kg)}}$$

The transfer coefficient represents the ratio of concentration of radioactive cesium in crops to the concentration of radioactive cesium in soil and is calculated for each crop. It is possible to estimate the concentration of cesium in crops by multiplying the transfer coefficient by the concentration of radioactive cesium in soil. The transfer coefficients of key foods have been reported as follows: rice from 0.00021 to 0.012 (Tsukada *et al.* 2002), green leafy vegetables from 0.00007 to 0.076, fruits from 0.0038 to 0.023 and root vegetables from 0.0008 to 0.36 (Ministry of Agriculture, Forestry and Fisheries 2011). The coefficients may vary quite widely because the transfer of radioactive cesium from soil to crops is greatly influenced by the nutrient conditions of soil and the physical state of cesium in soil.

Based on the transfer coefficient for rice, areas where rice planting was allowed in 2011 were required to have concentrations of radioactive cesium in soil under 5000 Bq/kg. It was believed that cultivation of soil under this limit would produce white rice under the provisional limit of 500 Bq/kg for radioactivity in food. In reality, of the 20,000 or more bags of whole rice tested, more than 95 percent were under 50 Bq/kg, and almost all the rest were under the provisional limits (Ministry of Agriculture, Forestry and Fisheries 2013). However, some whole rice containing radioactive cesium over the provisional limits was detected, and this discovery was widely reported by the media. The results of later investigative research showed that the rice containing high concentrations of radioactive cesium was produced in fields where concentrations of potassium in the soil were low and that, if sufficient potassium was applied to these fields, rice low in radioactive cesium could have been produced. In 2012 rice production was only begun after such measures had been taken and all the rice produced passed tests for radioactive cesium. The result of the tests for 2012 showed that of all the rice produced 99.95 percent was under 50 Bq/kg for concentrations of radioactive cesium.

Radioactivity in food and countermeasures

Provisional limits and safety criteria

Immediately following the Fukushima Daiichi nuclear accident, inspections of agricultural and fisheries products were initiated not only for Fukushima Prefecture but for all of Japan. These inspections identified that leafy vegetables from the northern Kanto and southern Tohoku regions contained high concentrations of iodine-131 and cesium-137. Accordingly, on March 17, 2011, the national government issued provisional limits for radioactivity in food, including less than 2000 Bq/kg for radioactive iodine and less than 500 Bq/kg for radioactive cesium. Shipment of products over these limits was prohibited.

The provisional limits were only an emergency action, and, due to criticism of the fact that these values were high in comparison with limits in other countries, on April 1, 2012 the Ministry of Health, Labor and Welfare set new limits for radioactive materials in food. These new limits set the criteria for radioactive cesium in food according to type of food product, including water (10 Bq/kg), general food products (100 Bq/kg) and milk and infant-food products (50 Bq/kg). These values were calculated based on guidelines from the ICRP that are meant to ensure that annual cumulative additional exposure remains below 1 mSv. Additionally, although it is believed that the amounts of strontium and plutonium released as a result of the Fukushima accident are limited, it will take considerable time and effort to determine and measure the amount of these substances present in the environment. Hence these policies err on the side of safety and set conservative limits (Ministry of Health, Labour and Welfare 2012).

Radioactivity in food and internal exposure

Currently, agricultural and fisheries products from Fukushima and neighboring prefectures are subject to inspection for radioactivity by various actors, including the national government, prefectural governments, local governments, local residents, and distributors and retailers. In addition, these inspections take place at various stages, including production, shipment, distribution, retail, and consumption. For example, in 2013, Fukushima Prefecture inspected 28,770 samples of 469 different products at the shipment stage and every package of rice, a staple food in Japan was inspected (Fukushima Prefecture 2013). In comparison with 2011, radionuclides in food have greatly decreased. Radioactive iodine, which has a short half-life, is now completely undetected in food products. For radioactive cesium, food products such as rice, vegetables, fruits, and livestock are nearly all under the 100 Bq/kg limits for radiation in food and, moreover, almost no radioactive cesium has been detected. However, in the case of wild mushrooms, wild edibles, wild game, and some marine products, foods over 100 Bq/kg have been identified, and their shipment continues to be restricted.

The "market basket" inspections conducted at the distribution stage have not detected any food products over 100 Bq/kg of radioactive cesium and the majority of products contain no radioactive cesium. Additionally, the *kagezen* inspections, in which every household prepares one extra meal which is comprehensively checked by a measuring device, conducted by the national government and consumer organizations, have only very rarely detected radioactive cesium—and the concentration of radioactive cesium in the atmosphere has also dropped to an extremely low level. Based on these results, the calculated level of internal exposure is now 2–3 orders of magnitude lower than exposure from natural background radiation. In regard to radioactive cesium persisting in the body, the many residents of Fukushima Prefecture subjected to whole-body testing have nearly all been clear of any radioactive cesium. Interviews with individuals who have been found to have radioactive cesium in their bodies reveal that all have regularly consumed wild mushrooms and other wild edibles that do not enter into formal distribution channels (Hayano *et al.* 2013).

Conclusion

In this chapter we attempted to trace the "invisible" processes of the diffusion and deposition of radioactive materials, emitted as a result of the nuclear accident at the Fukushima Daiichi NPP in March 2011 as a precondition of understanding the socio-economic, political, and cultural consequences of radiation contamination, and to gain insight into the prospects for recovery. We focused primarily on the diffusion and deposition of iodine-131, cesium-134, and cesium-137 because these were the most problematic materials in terms of their amount and effects on living organisms. We now have a relatively good understanding of the process and extent of the transfer of radioactive materials in atmospheric and terrestrial environments, at least for the initial period of the nuclear disaster. Nevertheless, the unavailability of reliable information, the national government's poor handling of the nuclear disaster during the early period, and the spread of nuclear contamination beyond the initial predictions made it difficult for the residents of Fukushima, as well as many Japanese people, to gain confidence in the information presented to them and in the authority itself. This broad lack of public confidence has had a critical consequence, for example, in securing food safety. The safety standards implemented after the nuclear accident are high, inspection systems are well established, and virtually no food in the market reveals higher than normal radiation. Yet consumers still tend to avoid food produced in Fukushima, and the food products are severely undervalued in the market. While these problems are socially and politically rooted, and need to be addressed accordingly, it is still imperative to conduct long-term monitoring of the movement of radioactive materials and to use the data to reduce radiation exposure by advancing decontamination measures that are effective and efficient.

Acknowledgment

This work was supported by JSPS KAKENHI Grant Number 25220403.

References

Chino, Masamichi, Hiromasa Nakayama, Haruyasu Nagai, Hiroaki Terada, Genki Katata, and Hiromi Yamazawa. 2011. "Preliminary Estimation of Release Amounts of ^{131}I and ^{137}Cs Accidentally Discharged from the Fukushima Daiichi Nuclear Power Plant into the Atmosphere." *Journal of Nuclear Science and Technology*, 48(7): 1129–a34.

De Cort, M. 1998. "Atlas of Caesium Deposition on Europe after the Chernobyl Accident." *The Radiological Consequences of the Chernobyl Accident*, Luxembourg: Office for Official Publications of the European Communities.

Draxler, Roland, Dèlia Arnold, Masamichi Chino, Stefano Galmarini, Matthew Hort, Andrew Jones, Susan Leadbetter, Alain Malo, Christian Maurer, Glenn Rolph, Kazuo Saito, René Servranckx, Toshiki Shimbori, Efisio Solazzo, and Gerhard Wotawa. 2015. "World Meteorological Organization's Model Simulations of the Radionuclide Dispersion and Deposition from the Fukushima Daiichi Nuclear Power Plant Accident." *Journal of Environmental Radioactivity*, 139: 172–84.

Evangeliou, Nikolaos, Yves Balkanski, Anne Cozic, and Anders Pape Møller. 2013. "Global Transport and Deposition of ^{137}Cs Following the Fukushima Nuclear Power Plant Accident in Japan: Emphasis on Europe and Asia Using High-Resolution Model Versions and Radiological Impact Assessment of the Human Population and the Environment Using Interactive Tools." *Environment Science and Technology*, 47(11): 5803–12.

Fukushima Prefecture. 2013. "Mizu Shokuhin tou no Houshasei Busshitsu Kensa" [The Radioactivity Measurements of Water and Foods]. Accessed October 5, 2015. www.pref.fukushima.lg.jp/site/portal/list280.html. [In Japanese.]

Fukushima Prefecture. 2014. "Kankyo Houshansen Monitoring, Mesh Chousa Kekka Jouhou" [Mesh Results of Environmental Radiation Monitoring]. Accessed October 5, 2015. www.pref.fukushima.lg.jp/sec/16025d/monitaring-mesh.html. [In Japanese.]

Hayano, Ryugo S., Masaharu Tsubokura, Makoto Miyazaki, Hideo Satou, Katsumi Sato, Shin Masaki, and Yu Sakuma. 2013. "Internal Radiocesium Contamination of Adults and Children in Fukushima 7 to 20 Months after the Fukushima NPP Accident as Measured by Extensive Whole-Body-Counter Surveys." *Proceedings of the Japan Academy Series B, Physical and Biological Sciences*, 89(4): 157–63.

Hong, G. H., M. A. Hernández-Ceballos, R. L. Lozano, Y. I. Kim, H. M. Lee, S. H. Kim, S. W. Yeh, J. P. Bolívar, and M. Baskaran. 2012. "Radioactive Impact in South Korea from the Damaged Nuclear Reactors in Fukushima: Evidence of Long and Short Range Transport." *Journal of Radiological Protection*, 32(4): 397–411.

Hsu, Shih-Chieh, Chih-An Huh, Chuen-Yu Chan, Shuen-Hsin Lin, Fei-Jan Lin, and Shaw Chen Liu. 2012. "Hemispheric Dispersion of Radioactive Plume Laced with Fission Nuclides from the Fukushima Nuclear Event." *Geophysical Research Letters*, 39(7): L00G22.

Huh, Chih-An, Shih-Chieh Hsu, and Chuan-Yao Lin. 2012. "Fukushima-derived Fission Nuclides Monitored around Taiwan: Free Tropospheric Versus Boundary Layer Transport." *Earth and Planetary Science Letters*, 319–20: 9–14.

Huh, Chih-An, Chuan-Yao Lin, and Shih-Chieh Hsu. 2013. "Regional Dispersal of Fukushima-derived Fission Nuclides by East-Asian Monsoon: A Synthesis and Review." *Aerosol and Air Quality Research*, 13(2): 537–44.

International Atomic Energy Agency (IAEA). 2006. *Environmental Consequences of the Chernobyl Accident and Their Remediation: Twenty Years of Experience*, Vienna: IAEA.

Japan Meteorological Agency. 2015. "Fukushima ken Hamadoori chihou Fuukou no Keikou zu (nen)" [Annual Trends of the Wind Direction in Central Fukushima]. Accessed October 5, 2015. www.data.jma.go.jp/obd/stats/etrn/wdr/windMap.php. [In Japanese.]

Katata, Genki, Hiroaki Terada, Haruyasu Nagai, and Masamichi Chino. 2012a. "Numerical Reconstruction of High Dose Rate Zones due to the Fukushima Dai-ichi Nuclear Power Plant Accident." *Journal of Environmental Radioactivity*, 111: 2–12.

Katata, Genki, Masakazu Ota, Hiroaki Terada, Masamichi Chino, and Haruyasu Nagai. 2012b. "Atmospheric Discharge and Dispersion of Radionuclides during the Fukushima Dai-ichi Nuclear Power Plant Accident. Part I: Source Term Estimation and Local-scale Atmospheric Dispersion in Early Phase of the Accident." *Journal of Environmental Radioactivity*, 109: 103–13.

Katata, G., M. Chino, T. Kobayashi, H. Terada, M. Ota, H. Nagai, M. Kajino, R. Draxler, M. C. Hort, A. Malo, T. Torii, and Y. Sanada. 2014. "Detailed Source Term Estimation of the Atmospheric Release for the Fukushima Daiichi Nuclear Power Station Accident by Coupling Simulations of Atmospheric Dispersion Model with Improved Deposition Scheme and Oceanic Dispersion Model." *Atmospheric Chemistry and Physics Discussions*, 15(2): 1029–70.

Kim, Chang-Kyu, Jong-In Byun, Jeong-Suk Chae, Hee-Yeoul Choi, Seok-Won Choi, Dae-Ji Kim, Yong-Jae Kim, Dong-Myung Lee, Won-Jong Park, Seong A Yim, and Ju-Yong Yun. 2012. "Radiological Impact in Korea following the Fukushima Nuclear Accident." *Journal of Environmental Radioactivity*, 111: 70–82.

Long, N. Q., Y. Truong, P. D. Hien, N. T. Binh, L. N. Sieu, T. V. Giap, and N. T. Phan. 2012. "Atmospheric Radionuclides from the Fukushima Dai-ichi Nuclear Reactor Accident Observed in Vietnam." *Journal of Environmental Radioactivity*, 111: 53–8.

Lozano, R. L., M. A. Hernández-Ceballos, J. A. Adame, M. Casas-Ruíz, M. Sorribas, E. G. San Miguel, and J. P. Bolívar. 2011. "Radioactive Impact of Fukushima Accident on the Iberian Peninsula: Evolution and Plume Previous Pathway." *Environment International*, 37(7): 1259–64.

Lujanienė, G., S. Byčenkienė, P. P. Povinec, and M. Gera. 2012. "Radionuclides from the Fukushima Accident in the Air over Lithuania: Measurement and Modelling Approaches." *Journal of Environmental Radioactivity*, 114: 71–80.

Masson, O., A. Baeza, J. Bieringer, K. Brudecki, S. Bucci, M. Cappai, F. P. Carvalho, *et al.* 2011. "Tracking of Airborne Radionuclides from the Damaged Fukushima Dai-Ichi Nuclear Reactors by European Networks." *Environmental Science and Technology*, 45(18): 7670–7.

Ministry of Agriculture, Forestry and Fisheries. 2011. *Nochi Dojo chu no Hoshasei Sesiumu no Yasairui oyobi Kajitsurui eno Iko no Teido* [Degrees of Transfer of Radioactive Cesium in Agricultural soil to vegetables and fruits], Tokyo: Ministry of Agriculture, Forestry and Fisheries. Accessed October 5, 2015. www.maff.go.jp/j/press/syouan/nouan/pdf/110527-01.pdf. [In Japanese.]

Ministry of Agriculture, Forestry and Fisheries. 2013. *Heisei 24 Nenndo made no Nosan Butsu ni fukumareru Hoshasei Sesiumu no Nodo no Kensa Kekka no Gaiyo* [An Overview of Inspection Results of Radioactive Cesium Concentration in Agricultural Products until Heisei 24 (2012)], Tokyo: Ministry of Agriculture, Forestry and Fisheries. Accessed October 5, 2015. www.maff.go.jp/j/kanbo/joho/saigai/s_chosa/H24gaiyou.html#kome. [In Japanese.]

Ministry of Environment. 2013. *Progress on Off-site Cleanup Efforts in Japan, July 17*, Tokyo: Ministry of Environment. Accessed October 5, 2015. http://josen.env.go.jp/en/pdf/progressseet_progress_on_cleanup_efforts.pdf?141022.

Ministry of Health, Labour and Welfare. 2012. *Shokuhin chu no Houshasei Busshitsu no Taisaku to Genjou ni tsuite* [*The Countermeasures and Current Situation of the Radioactivity in Foods*], Tokyo: Ministry of Health, Labour and Welfare. Accessed October 5, 2015. www.mhlw.go.jp/shinsai_jouhou/dl/20131025-1.pdf. [In Japanese.]

Morino, Yu, Toshimasa Ohara, and Masato Nishizawa. 2011. "Atmospheric Behavior, Deposition, and Budget of Radioactive Materials from the Fukushima Daiichi Nuclear Power Plant in March 2011." *Geophysical Research Letters*, 38(7): L00G11.

Morino, Yu, Toshimasa Ohara, Mirai Watanabe, Seiji Hayashi, and Masato Nishizawa. 2013. "Episode Analysis of Deposition of Radiocesium from the Fukushima Daiichi Nuclear Power Plant Accident." *Environmental Science and Technology*, 47(5): 2314–22.

Nagai, Haruyasu, Genki Katata, Hiroaki Terada, and Masamichi Chino. 2014. "Source Term Estimation of ^{131}I and ^{137}Cs Discharged from the Fukushima Daiichi Nuclear Power Plant into the Atmosphere." In *Radiation Monitoring and Dose Estimation of the Fukushima Nuclear Accident*, edited by Sentaro Takahashi, 155–73. Tokyo: Springer.

Nuclear Safety Commission. 2011. "The Evaluation about the Furnace State of the First, Second and Third Units Concerned with an Accident of the Tokyo Electric Power Fukushima First Nuclear Power Plant Co., Ltd." Accessed October 5, 2015. www.meti.go.jp/press/2011/06/20110606008/20110606008-2.pdf. [In Japanese.]

Ohara, Toshimasa, Yu Morino, and Atsushi Tanaka. 2011. "Atmospheric Behavior of Radioactive Materials from Fukushima Daiichi Nuclear Power Plant." *Japan National Institute of Public Health*, 60(4): 292–9. [In Japanese.]

Ohse, Kenji, Nobuharu Kihou, Katsuaki Kurishima, Yasushi Fukuzono, and Ichiro Taniyama. 2012. "Influence of Land Use for Radiocesium Content in Soil before and after Fukushima Daiichi Nuclear Power Plant Accident." Annual Meeting, Japanese Society of Soil and Plant Nutrition, Tottori.

Ohse, Kenji, Nobuharu Kihou, Katsuaki Kurishima, Tsunehisa Inoue, and Ichiro Taniyama. 2015. "Changes in Concentrations of 131I, 134Cs and 137Cs in Leafy Vegetables, Soil and Precipitation in Tsukuba City, Ibaraki, Japan, in the First 4 Months after the Fukushima Daiichi Nuclear Power Plant Accident." *Soil Science and Plant Nutrition*, 61(2): 225–9.

Sanami, Toshiya, Shinichi Sasaki, Kazuhiko Iijima, Yuji Kishimoto, and Kiwamu Saito. 2011. "Time Variations in Dose Rate and γ Spectrum Measured at Tsukuba City, Ibaraki, due to the Accident of Fukushima Daiichi Nuclear Power Station." *Transactions of the Atomic Energy Society of Japan*, 10(3): 163–9. [In Japanese.]

Stohl, A., P. Seibert, G. Wotawa, D. Arnold, J. F. Burkhart, S. Eckhardt, C. Tapia, A. Vargas, and T. J. Yasunari. 2012. "Xenon-133 and Caesium-137 Releases into the Atmosphere from the Fukushima Dai-ichi Nuclear Power Plant: Determination of the Source Term, Atmospheric Dispersion, and Deposition." *Atmospheric Chemistry and Physics*, 12(5): 2313–43.

Terada, Hiroaki, Genki Katata, Masamichi Chino, and Haruyasu Nagai. 2012. "Atmospheric Discharge and Dispersion of Radionuclides during the Fukushima Dai-ichi Nuclear Power Plant Accident. Part II: Verification of the Source Term and Analysis of Regional-scale Atmospheric Dispersion." *Journal of Environmental Radioactivity*, 112: 141–54.

Tsukada, H., H. Hasegawa, S. Hisamatsu, and S. Yamasaki. 2002. "Rice Uptake and Distributions of Radioactive 137Cs, Stable 133Cs and K From Soil." *Environmental Pollution*, 117(3): 403–9.

United States Environmental Protection Agency (US EPA). 2012. "RadNet Laboratory Data." Accessed October 5, 2015. www2.epa.gov/radnet/2011-japanese-nuclear-incident.

Winiarek, Victor, Marc Bocquet, Olivier Saunier, and Anne Mathieu. 2012. "Estimation of Errors in the Inverse Modeling of Accidental Release of Atmospheric Pollutant: Application to the Reconstruction of the Cesium-137 and Iodine-131 Source Terms from the Fukushima Daiichi Power Plant." *Journal of Geophysical Research*, 117: D05122.

3 Place stigmatization through geographic miscommunication

Fallout of the Fukushima nuclear accident

Takashi Oda

Introduction

The massive earthquake and tsunami of March 11, 2011, triggered a severe accident at the Tokyo Electric Power Company's (TEPCO) Fukushima Daiichi Nuclear Power Plant (NPP). Following the hydrogen explosions at the plant and the subsequent release of radioactive materials into the air, the central government issued a series of announcements and directives to evacuate, which were communicated to the public through numerous media outlets.

Due to the rarity of nuclear accidents of this magnitude, effective risk communication became a major challenge. In essence, partial and inaccurate geographic information presented by mass media during the period of successive evacuation announcements stigmatized[1] the cities of Iwaki in the south and Minami-Soma in the north. This chapter examines the process of information dissemination in the wake of the nuclear disaster, with a particular focus on the miscommunication of geographic information and its consequences by focusing on the case of Iwaki City, Fukushima Prefecture. This study draws on records of media coverage during the disaster, some of which were on the internet and some of which were derived from interviews with individuals with first-hand experience of the disaster. Based on the findings, I offer some policy suggestions intended to improve future emergency responses.

Risk studies and stigma

Risk is intimately related to "how society values the future, nature, and human well-being" (Kasperson and Kasperson 1996, 104), and it is eminently political because it concerns questions about whose values are more important. There is growing social-scientific interest in the ways that media reports influence people's perceptions of risks, particularly risks related to natural and environmental hazards. Studies often point out that experts, such as scientists and policy makers, are different from ordinary citizens in their perceptions of risks, which may lead to conflicts. These conflicts may be particularly acute when "citizen experts" (Tesh 1999), equipped with expert-like knowledge and skills, emerge, articulate the ways that a particular phenomenon is safe or dangerous, and actively participate in environmental policymaking discussions. Mediating and, in some cases, amplifying these conflicts are the mass media, which play pivotal parts in shaping public perceptions of risks (Wakefield and Elliot 2003). Indeed, Cowan *et al.* (2002) suggest that, although people are usually aware that news media tend to be selective and sensational, they often do not account for these characteristics when they evaluate and act on information reported about major events. Thus, mass media are vital to the potential production of stigma associated with natural and man-made disasters.

In their discussion of the social-amplification-of-risk framework, Kasperson and Kasperson (2005, 172) define stigma as "a mark placed on a person, place, technology, or product, associated with a particular attribute that identifies it as different and deviant, flawed or undesirable," arguing that the formation of stigma (i.e. stigmatization) involves three stages, as follows. First, through communication processes, risk-related attributes are given high visibility, tied to impressions created by images of high riskiness, leading to the social amplification of risk. Second, marks are used to identify particular persons, places, technologies, or products as risky and undesirable. Last, the social amplifications of risk and the marking change the social identities of the marked persons, places, technologies, or products, which leads to behavioral changes by the individuals, places, entities, or things that are marked (Kasperson and Kasperson, 2005, 171).

Numerous empirical case studies on stigmatization have been conducted, led by the Kaspersons' contributions to the study of the social dimensions of risks since the 1980s. For example, empirical accounts in *The Social Contours of Risk Volume I: Publics, Risk Communication and the Social Amplification of Risk* (Kasperson and Kasperson 2005) explicate the ways that risks are communicated to the public and to stakeholders, amplified and distorted through the media, and how societies respond to them. These studies point out the unequal distribution of risk in a variety of contexts, such as facility-siting, hazardous-waste management, and global climate change. However, few studies have dealt with the ways that actual location-based risk-information dissemination, such as place-name reporting and mapping, has influenced the processes and outcomes of stigmatization after catastrophes.[2] The present case study of the evacuation process of the Fukushima nuclear accident illustrates this process, drawing on the social-amplification-of risk-framework.

Stigmatization through geographic miscommunication is a crucial issue for policymaker, particularly in the areas of development and disaster risk reduction (DRR), because of the increasing attention paid to the value of using location-based risk information. Because of the post-Hyogo Framework for Action (HFA) agreement, DRR practitioners in the world recently adopted the Sendai Framework for Disaster Risk Reduction 2015–2030 (UNISDR 2015); and the outcome document of the third World Conference on Disaster Risk Reduction (WCDRR) in Sendai, Japan, clearly stated in its action priories that it is important to "develop, periodically update and disseminate, as appropriate, location-based disaster risk information, including risk maps, to decision makers, the general public and communities at risk of exposure to disaster in an appropriate format by using, as applicable, geospatial information technology," and to "promote real time access to reliable data, make use of space and in situ information, including geographic information systems (GIS)" (UNISDR 2015, 12).

The Sendai Framework may foster the use of geospatial technologies in the implementation of DRR and emergency management worldwide. Accordingly, professionals in mass communications should be aware of the effects of their uses of geographic information in risk communication on people's reactions to and decisions regarding first responses and evacuation in catastrophic situations. This case study therefore contributes to future discussions that address these recent developments.

Place stigmatization as fallout of the nuclear accident: the case of Iwaki City

Iwaki City is located in the southeastern corner of Fukushima Prefecture, bordering Ibaraki Prefecture. It has a long coastline and has been severely affected by both the earthquake and tsunami of March 11, 2011. The earthquake and the tsunami inundated the coastal communities in Iwaki, and many houses were destroyed or rendered uninhabitable and condemned. In Shimogawa, Iwaki Sun Marina was hit by multiple high waves that destroyed the buildings, berths, and related facilities. About 150 vessels that had been moored at the harbor were lost to the sea. Many of the facilities at the Port of Onahama, one of the major international trading and fishing ports in eastern Japan, were also destroyed. Fishing vessels that had been moored at the port were carried several miles inland by the tsunami. According to the Iwaki Disaster Management Headquarters, there were 310 casualties, 37 people missing, and 88,696 damaged buildings in Iwaki City.

Despite these damages from the earthquake and tsunami, Iwaki faced the distinct challenge of acute stigmatization associated with the nuclear accident during a critical period of the disaster. In short, the news media's and the government's miscommunications of geographic information in its evacuation orders to the public increased the public's panic and significantly delayed Iwaki's receipt of emergency aid and, thereby, its recovery. In the first stage of stigmatization, risky images are attributed, in this case to the place name(s), via mass media, such as television, radio, and/or newspaper. Moreover, individuals who are misinformed when fear levels are high may misquote statements from internet social-networking sites. The images marked the Iwaki as risky and therefore undesirable, which damaged aspects of the economy, such as tourism and food production (for example, agriculture and fisheries), because public attitudes were altered toward things attributed to the places. Although the mass media and gossip have some responsibility for the spread of misinformation, governments have the capacity to counter misinformation and manage the dissemination of news, and they can be the principle sources of risk information. Governments can ensure the spread of accurate information and prevent stigmatization.

Risk communication and stigma in the making

The stigma attached to the disaster-afflicted areas affected the public after the earthquake and NPP accident through the media coverage of the events as they unfolded. On March 12, 2011, the day after the earthquake, households in the tsunami-affected areas not designated as evacuation zones began voluntary evacuations, particularly households with small children. Public anxiety increased as a series of explosions and fires were reported. No explanations were forthcoming from either the government or the TEPCO management, while the televised video images of the explosion and the commentaries by invited nuclear experts created fear in the viewers' minds.

Three days later, on March 15, 2011, at 11:00 am, the Japanese government issued the order for all people living inside a 20 km radius of the Fukushima Daiichi NPP to evacuate, and for people living between 20 km and 30 km from the NPP to stay indoors. Prime Minister Naoto Kan gave a televised message to the country regarding the worsening situation at the Fukushima Daiichi NPP and the increasing risk caused by the spread of radioactive materials, stating, in part, that,

> in view of the developing situation, those who are outside the 20-km radius but still [inside] a 30 km radius should remain indoors in their house, office, or other structure, and not go outside. Further, with regard to the Fukushima Daini Nuclear Power Station, most people have already evacuated beyond a 10-km radius but we are calling for everyone who remains [inside] that radius to fully evacuate to a point beyond it.[3]

Specifically, given the grave situation, he asked all people living between 20 km and 30 km from the NPP to "please not go out and stay inside the house or offices." However, Prime Minister Kan did not specifically name the locations targeted for evacuation or sheltering indoors. Chief Cabinet Secretary Yukio Edano repeated the information about the 20 km to 30 km precautions and, this time, stated the names of the municipalities inside the 20 km to 30 km radius. Edano also noted that some of the municipalities on his list were not entirely inside the zone, that other municipalities overlapped the 20 km zones, and that only the northern portion of Iwaki was included in the 30 km zone. NHK, Japan's public broadcasting organization, superimposed bulletin messages on its live broadcasts that read, "20 km–30 km also for sheltering indoors" (Fukunaga 2011).

Following the March 15 indoor-sheltering order, television media reported that, as a result of the extension of the evacuation zone to 30 km, including the areas of the indoor-sheltering order, part of Iwaki was now inside the designated zone.[4] Updated at 12:39 pm on March 15, NHK news online issued a version entitled "Iwaki City and Iitate Village subject to indoor sheltering." The news script, which NHK archived in its online historical records, can be translated into English as, "Chief Cabinet Secretary Edano, at a press conference, revealed that Iwaki City and Iitate Village [to the northwest of the NPP] are added as municipalities subject to sheltering indoors to the existing 10 municipalities subject to the evacuation order." Following the news statement notifying the public of the 10 municipalities already included in the evacuation order, NHK news announced that, "within the 30 km radius, in addition to these 10 municipalities, Iwaki City and Iitate Village are newly added." The major television news networks similarly reported the extension of the security zone to a 30-km radius.

The map on NHK television news on March 15 showed only the northern tip of Iwaki City (which touched the 30 km radius), despite the city's vast area, because the majority of the populated municipal area was deleted from the image to fit the television screen aspect ratio. Copyright protection prohibits my presentation of the actual map used by NHK, but Figure 3.1 shows the area depicted by NHK with the same aspect ratio.[5] The figure suggests that a large proportion of Iwaki was excluded from the news map and that the televised map was not helpful to people outside of Fukushima for understanding Iwaki's size and relationships to other municipalities and prefectures. Neither the map nor the way that the information was disseminated provided the viewers with focused information about the areas in the 30 km indoor-sheltering zone.

The NHK reports gave the impression that the entire city of Iwaki, with a population of about 342,000 in 2010,[6] had been designated as dangerous and must be evacuated. Thus began the place-based stigmatization that affected Iwaki (and, likely, all of Fukushima Prefecture). A municipal official of Iwaki City interviewed by the accident investigation board recalled that several citizens telephoned the city and asked whether NHK's reports that the entire city of Iwaki was in the indoor-sheltering zone were true or not.[7]

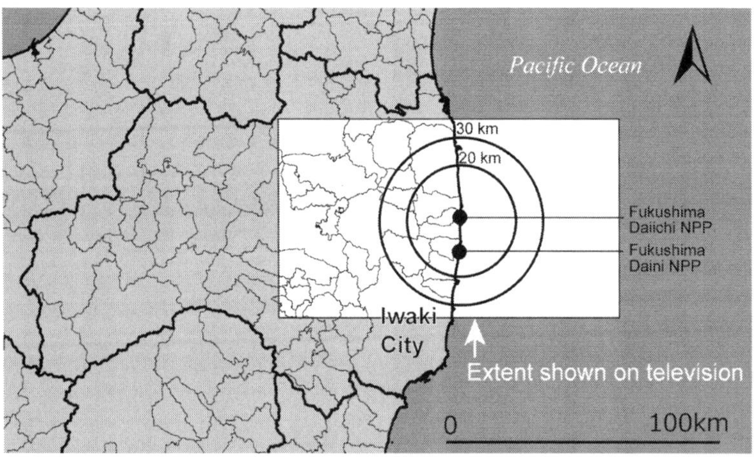

Figure 3.1 Areas of Fukushima shown on the NHK news in the context of the wider region.

History and geography of Iwaki City

The city of Iwaki was incorporated on October 1, 1966, in one of the first national municipal mergers. It was established by combining 14 smaller municipalities to create a Core City, which is a Japanese local governmental designation.[8] Iwaki was Japan's largest municipality in land area (at 1,231.13 km^2, it was almost twice as large as the 23-ward Tokyo Metropolitan Area) until April 1, 2003, when Shizuoka merged with Shimizu to create the newly defined Shizuoka City. Iwaki's northern border is approximately 30 km from the Fukushima Daiichi NPP, and its farthest border is about 60 km from the NPP (Figure 3.2).

Figure 3.2 compares the municipal boundaries of the eastern part of Fukushima in 1920 to those of 2011. Both maps show the radial distance from the present Fukushima Daiichi NPP location. The 2011 map illustrates that the 30 km indoor-sheltering zone barely extended into Iwaki municipality.

Despite the geographic facts, however, the public messages regarding health risks and restrictions were miscommunicated and lacked precise geographic information and explanation. The result was a public perception that the entire city of Iwaki was part of the nuclear restricted zone, along with the attributed image of Fukushima Prefecture as risky. This misperception triggered a serious aspect of stigmatization, because the actual geographic extent of this huge municipality of over 340,000 residents, once the largest municipality in Japan, was not well known outside of Fukushima.

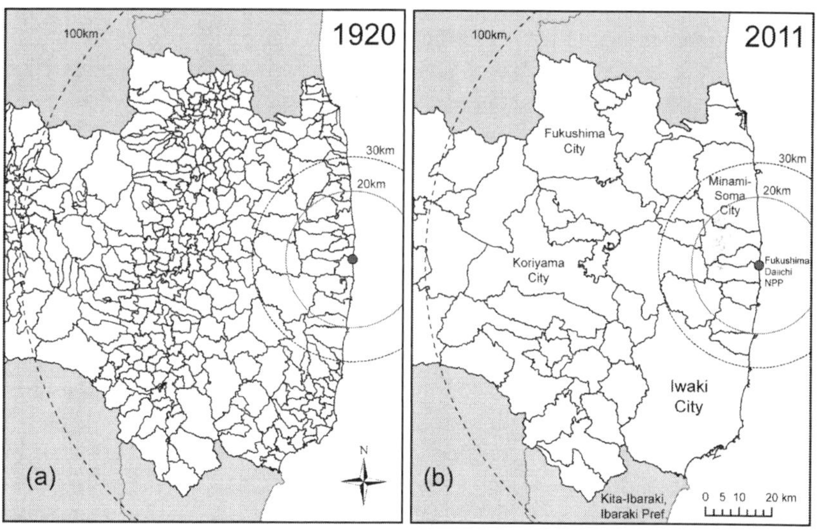

Figure 3.2 Municipal boundaries in Fukushima in (a) 1920 and (b) 2011.

Source: The author, with data from the National Land Numerical Information, Ministry of Land, Infrastructure, Transport and Tourism website at http://nlftp.mlit.go.jp/ksj/.

Outcome of stigmatization as changed public behavior

As the perception that all of Iwaki was dangerous spread throughout the public, private fleet companies refused to enter the city, even to destinations that were 60 km outside the 30 km security zone. For example, some truck drivers coming from Kita Ibaraki, Ibaraki Prefecture, turned around when they reached the southern end of Iwaki City. Similar incidents were reported in Minami-Soma, whose southern area was in the evacuation and indoor-sheltering zones. Minami-Soma City was a relatively young municipality, amalgamated in 2006 with the city of Haramachi (into which the towns of Odaka and Kashima had been merged). The stigma-oriented shunning prompted Mayor Katsunobu Sakurai of Minami-Soma to plead for help on television and upload a YouTube video to communicate in Japanese with English subtitles. By the end of April, more than 110,000 views of his plea had been logged.

In a response to similar calls by local leaders, Chief Cabinet Secretary Yukio Edano reported in a press conference on March 16, 2011, that special requests had been made by the governor of Fukushima Prefecture for help with problems regarding private aid-delivery companies that were avoiding areas such as Iwaki and Minami-Soma. Mr Edano stated:

> There has been a great deal of overreaction… We have received reports that a situation is arising in which, chiefly in the private sector, supplies are not reaching people in some areas, even though these areas lie outside the area [where] we have instructed people to remain indoors. I want you to understand that even inside the area where we have called for people to remain indoors, the levels [of radiation] are not sufficient to have any immediate effect on the health of someone carrying out activities outdoors, even inside this area… The majority of Iwaki City, for example, lies in an area more than 30 km away from the site, and I ask you please to continue to carry out regular shipments and distribution[s] to people in these areas.[9]

The environmental radiation levels at the time of Edano's press conference (starting at 5:56 pm on March 16, 2011) are shown in Figure 3.3. They read 1.74 μSv/h in central Iwaki (Taira area), 3.63 μSv/h in Minami-Soma, 3.30 μSv/h in Shirakawa, 2.94 μSv/h in Koriyama, and 14.60 μSv/h in City of Fukushima.[10] Given the direction of the radioactive plume (Chapter 2), Iwaki's radiation level was one of the lowest among major cities in Hama-Dori and Naka-Dori. Although what these levels mean to the extent or type of health risks is still a matter of scientific debate, and the accuracy and validity of these measurements could be questioned,[11] it is clear in retrospect that the perception of all of Iwaki being heavily contaminated was misguided.

Nonetheless, emergency-supplies workers did not have a clear understanding of the geographical spread of radioactive materials at that time. Consequently, truck drivers delivering emergency aid supplies through some areas, such as Koriyama and Fukushima cities, which were later reported to have higher levels of radiation, turned around at the northern border of Iwaki, whose northern fringe

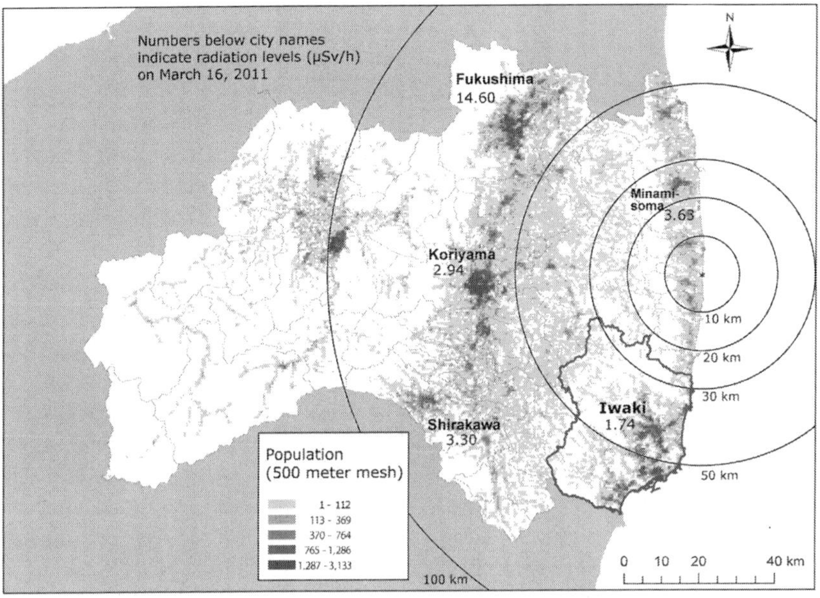

Figure 3.3 Population (2010) in Fukushima Prefecture and radii distances from the Fukushima Daiichi NPP.

Source: The author, with the 2010 census regional mesh statistics, Ministry of Home Affairs.

Note: Figures below the municipal names are the levels of radiation (μSv/h), measured at the specified sites in the respective municipalities.

was designated as an indoor-sheltering zone. A similar situation was reported at the southern Iwaki border; gasoline trolleys refused to enter the southern part of Iwaki, located about 50–60 km from the Fukushima Daiichi NPP. It was later learned that major businesses and media organizations with branch offices in Iwaki had issued independent no-entry orders and evacuation orders for their workers in Iwaki. Some Iwaki City officials were compelled to use their personal drivers to transport vital supplies. Outsiders' reactions to the stigmatized place were more extreme; incidents were reported in which fears of contamination motivated people to deny cars with Iwaki license plates entry to gasoline stations, and people from Iwaki were asked to leave a neighborhood in the Tokyo Metropolitan Area to which they had been evacuated.

Had those who were in a position to control the creation and flow of information related to the evacuation zones been more sensitive to the historical and geographical realities of Iwaki City (such as the fact of the extensive merger), a more cautious approach when reporting evacuation warnings might have been taken. The failure to do so meant geographically sensitive information was poorly communicated, which resulted in the acute stigmatization of Iwaki City.

Mediating the risk perception: amended television coverage

Television and radio news are always significant in post-disaster risk communications. In Japan, the 1923 Great Kanto Earthquake led to the 1925 creation of AM radio broadcasting in the wake of rumors about Koreans and Socialists. During the 1923 disaster, the postal and newspaper services were disrupted, and the majority of the survivors had no information at all, which created public anxiety. Since then, television and radio broadcasting in Japan has developed into a major source of information for citizens and professionals in times of disaster. Even though other means of communication, including social-networking sites, such as Twitter and Facebook, are growing, television still remains a critical means to communicate important geographic-risk information to the public.

However, it was not until March 21, 2011, ten full days after the earthquake, that the televised news media began reporting the conditions in Iwaki. TV Asahi's morning television interview with a city-assembly member, Kazuyoshi Sato, covered the extent of the damages to the coastal communities, including the Port of Onahama (Fukushima's largest port) and the delays to recovery caused by the fears that arose from the misreporting of the nuclear fallout. The broadcast presented a map similar to Figure 3.3 that charted the distance between Fukushima Daiichi and Iwaki City. The reporter stressed that, although Iwaki City is the most extensive municipality in Fukushima Prefecture, and the 30 km security zone covered a very small part of the city, "delivery companies are so afraid to go into Iwaki that aid supplies and gasoline are rarely brought."[12] The reporter also stated that residents living outside the evacuation zones were voluntarily evacuating to Kita-Ibaraki due to the spreading fear of radiation. Other announcements included information about gasoline shortages, which were preventing locals from traveling or evacuating, a shortage of doctors and medical personnel in public hospitals, shortages of supplies and food, and cases of near starvation among both patients and staff in senior care.

Naruhito Iguchi, the lead reporter of the television program, displayed a map of the whole of Fukushima Prefecture, emphasizing the size of Iwaki City. The map conveyed the radiation level in Iwaki and superimposed radiation levels over the studio map. It showed that, as of March 21, 2011, radiation levels in Iwaki (0.73 μSv/h) were lower than in the cities of Fukushima (7.94 μSv/h) and Koriyama (2.54 μSv/h). Iguchi went on to say that emergency supplies were being delivered to the Naka-Dori area but not to Iwaki, and he stated that "drivers that came to Koriyama don't want to go to Iwaki, so people from Iwaki have to come pick stuff up." Iguchi explained that the cause of people's misunderstanding was rooted in the fact that the name "Iwaki" made a stronger impression on people's minds than the specific information that the area of interest was only the northern border area. Therefore, "Iwaki" was a stigmatizing concept, and the entire place was marked. The public had to somehow manage its fears of the unknown and the uncertain. It had no option but to react to the simple labels reported in the news and public announcements. Much was unknown, what was known was not clearly communicated, and the risky and fearful images that accompanied information about risks were amplified among the public.

After the reporter's presentation, the then mayor of Iwaki, Takao Watanabe, telephoned the reporter on live television. He stated that essential goods and supplies, such as water, food, gasoline, and drugs, were not being delivered at all, that the central government was giving no explanation of the problem, and that Iwaki City was actually taking in evacuees from Futaba District inside the nuclear evacuation zones. Mr Watanabe went on to say that engineers and plumbing professionals would have come from all over Japan to help rebuild the water-supply system, which had been damaged by the earthquake and tsunami. However, because of the public's attitude toward the stigmatized place, these resources were not being provided to Iwaki. The only staff available to repair the water system were those of the local water departments and the local private plumbing companies, and they had been working around the clock ever since the earthquake and tsunami had hit. It took much longer in Iwaki to repair the water system than in other cities that were similarly affected by the disaster due to its large geographic area and the lack of external resources.

On the afternoon of March 21, 2011, "Information Live Miyaneya," on Yomiuri News Network, reported similar issues in Iwaki.[13] That report, entitled "The surviving cities that were shunned," included maps similar to the one from the Asahi morning show overlaying the radiation levels, and it also included an interview with Mayor Watanabe. That same evening, Hodo Station, on TV Asahi, reported shortages of supplies and human aid workers and made a call for help based on an Iwaki viewer's email.

On March 22, 2011, the problems arising from stigma burdening Iwaki continued to be reported on the morning television shows. New reports described problems, such as the scarcity of heavy-equipment operators in Iwaki, the delayed removal of debris, and the slow rescue and recovery of missing people. One report quoted a local volunteer firefighter engaged in the recovery as saying, "one more week will make it much more difficult to identify dead bodies."[14] On the following day, an early morning news report on NHK was entitled, "Supplies not delivered even outside 30 km security zone."[15] It outlined the deteriorating conditions in Iwaki, including the fact that Iwaki City had begun to distribute supplies originally intended for shelter evacuees to some non-evacuee residents for fear that they would starve.

Conclusions

The place-based stigma imposed on Iwaki wreaked havoc. As in other instances of stigmatization, credible information about the risks was not available and the ways that the government delivered information did not reassure citizens about its validity, which led to anxieties, fears, and even panic. Indeed, the ways in which information was shared actually increased distrust of the government and TEPCO. Without clear and convincing information on which to base their behavior regarding Iwaki, people changed their attitudes and behaviors toward the stigmatized place, which slowed rescue, recovery, and infrastructure-repair efforts. The miscommunication through mass media amplified the risk perceptions. Although television broadcasting has sophisticated technology with which to display information on hazardous risks (for example, meteorological weather forecasting), it did not provide appropriately detailed information with visual mapping technology when it began reporting on March 15 about the extension of the restricted zones.

In addition to the problems related to the dissemination of information, there was the problem of the nature of the information itself. Radiation's invisibility and scientific complexity, its impacts on health, the terminology used to convey information about it, and the units and statistics (such as sievert (mSv, μSv), becquerel (Bq), and gray (Gy)), were far beyond the knowledge base of the average citizen and of those who reported on them in the media. Lack of confidence in reporters and other information sources intensified people's anxieties as television broadcasts persisted in presenting the worst possible spin on the situation. Inaccurate mapping of the risk areas, of the locations of municipalities with specific levels of risk, and of areas that needed assistance worsened the situation.

There is no intention here to downplay more direct damage by the disaster, such as serious radioactive contamination and the challenges that evacuees face, discussed in other chapters, but it is vitally important to consider the geographic perspective on what could have been done differently and what could be done to avoid a repetition of these problems in the event of a future disaster. First, official sources of information, which were, in this case, the government and TEPCO, must provide the public with immediate and clear risk information accompanied by visual spatial data, instead of stating, without explanation, "no immediate risks expected." In the case examined here, vague assurances did nothing more than amplify public anxiety regarding the honesty and transparency of the entities and people in positions of power. Telling people and, in particular, showing people maps pinpointing exactly where the potential dangers and hazards were so that these areas might be avoided would have eased anxiety. Providing visual comparisons of conditions and radiation levels that clarified the similarities to and differences from the normal conditions would have provided a risk context with which the public could have made informed, not stigmatizing, decisions.

Second, it is evident from people's reactions to city and other local names that simply stating the names of places considered safe or unsafe was insufficient to convey accurate understanding. Marking a stigmatized image on a place can increase bias toward or against a place based on the cultural or customary meanings given to the name. However, the geographic distances, locations, topography, demographics, and their interrelationships can clearly be explained by drawing on the existing scientific database and, in the case of the earthquake, they should have been explained, with supporting visual information, to communicate with the best possible accuracy the areas to avoid.

Third, one result of poor information dissemination during a disaster is individual and collective amplification of risk fears. Mapping the locations of relevant and vital resources for people's wellbeing would greatly assist their decision-making. Careful explanations that guide the readers of thematic maps, whether in newspapers or on television, would help people to understand the structure of rescue and support efforts and help them to make good decisions to protect their own and others' lives. Thus, sound geographic knowledge is vital in assisting and educating those in emergency management and in the media, particularly now that the Sendai Framework for Disaster Risk Reduction 2015–2030 clearly recognizes geospatial technology and location-based information as important tools for achieving priority action goals among DRR practitioners around the world.

Consumers shun Fukushima-made products. Thus, the effects of stigmatization linger on. The Kaspersons argued that stigma marks people, places, and things as flawed and undesirable, and this analysis has found that stigma indeed caused damage to Iwaki, Fukushima, and their people by miscommunication of geographic information and ungrounded rumor. Five years after the nuclear accident, acute stigmatization of Iwaki may be no longer evident, but broad stigmatization of Fukushima still lingers on; some people continue to avoid all things related to Fukushima. Indeed, it remains the major source of affliction for Fukushima's survivors of the triple disasters.

Acknowledgments

I would like to express deep gratitude to all those who have supported me throughout this research including family and friends in Iwaki, Fukushima, the editors, academic advisors, and many graduate students who provided insights and technical assistance. Its publication is funded by FY2013 Grant-in-Aid for Scientific Research (C) Number 25370911 of the Japan Society of the Promotion of Science (JSPS). Part of the discussion in this paper is based on a bulletin article in Japanese published online at Tohoku Geographical Association's special website in April 2011.

Notes

1 In this chapter, the term "stigma" refers to the equivalent Japanese term *fuhyo higai*, which means "damage (higai) caused by ungrounded rumor (fuhyo)." It is sometimes translated into English as reputational damage.

2 Sekine (2015) recently examined the ways that information about radiation after the Fukushima nuclear accident was reported by newspapers, other mass media, and in municipal announcements to people needing information, particularly the ways in which maps were used.

3 Prime Minister's Office: http://japan.kantei.go.jp/incident/110315_1100.html (Accessed February 22, 2015).

4 NHK: http://www3.nhk.or.jp/news/genpatsu-fukushima/index_0315.html (Accessed April 7, 2011).

5 The actual NHK map included superimposed names of the municipalities whose territories are fully or partially included in the 30 km radius, and the 10 km radius circle from Fukushima Daini (F2) NPP was shown to indicate the then security zones, which were later lifted.

6 City of Iwaki website: www.city.iwaki.fukushima.jp/dbps_data/_material_/localhost/01_gyosei/0110/H22kokusei-kakutei.pdf (Accessed February 22, 2015).

7 Cabinet Secretariat (2011): www.cas.go.jp/jp/genpatsujiko/hearing_koukai_4/024_025_koukai.pdf (Accessed June 30, 2016).
8 Article 252, Section 22 of the Local Autonomy Law of Japan.
9 Prime Minister's Office: http://japan.kantei.go.jp/incident/110316_1756.html (Accessed February 11, 2015).
10 Fukushima Prefectural Government measurements released at 6:00 pm, March 16, 2011.
11 JAEA (2011): Microsieverts and Other Nuclear-Related Units of Measure. www.jaea.go.jp/english/jishin/kaisetsu01.pdf (Accessed June 30, 2016).
12 March 21, 2011 TV Asahi at 8:48 am: "Super Morning" reporter Naruhito Iguchi.
13 March 21, 2011 Yomiuri TV: "Information Live Miyaneya" reporter Masatoshi Nakayama.
14 March 22, 2011 TV Asahi at 8:11 am: "Super Morning."
15 March 23, 2011 NHK General "Ohayo Nippon" at 7:20 am: reporter Fujinoki.

References

Cowan, Jodie, John McClure, and Marc Wilson. 2002. "What a Difference a Year Makes: How Immediate and Anniversary Reports Influence Judgments about Earthquakes." *Asian Journal of Social Psychology*, 5(3):169–85.

Fukunaga, Hidehiko. 2011. "Nuclear Disaster and Evacuation Information, the Media: The Case of Fukushima Dai-ichi Nuclear Power Plant Accident." *NHK Broadcasting Studies*, 61(9): 2–17. [In Japanese.]

Kasperson, Roger E. and Jeanne X. Kasperson. 1996. "The Social Amplification and Attenuation of Risk." *Annals of the American Academy of Political and Social Science*, 545(1): 95–105.

Kasperson, Roger E. and Jeanne X. Kasperson. eds. 2005. *Stigma and the Social Amplification of Risk: Towards a Framework of Analysis*, London: Earthscan.

Sekine, Ryohei. 2015. "How Was 'Radiation' Reported to the 'Local' People? An Examination through Information Dissemination from Municipal Government and the Media Coverage." In *Records of the Victims' Refuge Lives in the Great East Japan Earthquake*, edited by Naoki Yoshihara, Nihei Yoshiaki, and Michimasa Matsumoto, 687–714. Tokyo: Rikka Shuppan. [In Japanese.]

Tesh, Sylvia N. 1999. "Citizen Experts in Environmental Risk." *Policy Sciences*, 32(1): 39–58.

UNISDR. 2015. *Sendai Framework for Disaster Risk Reduction 2015–2030*. Accessed February 22, 2015. www.preventionweb.net/files/43291_sendaiframeworkfordrren.pdf.

Wakefield, Sara E. L., and Susan J. Elliot. 2003. "Constructing the News: the Role of Local Newspapers in Environmental Risk Communication." *The Professional Geographer*, 55(2): 216–26.

4 Living in suspension

Conditions and prospects of evacuees from the eight municipalities of Futaba District

Mitsuo Yamakawa

Introduction

The Great East Japan Earthquake Disaster was caused by an earthquake and tsunami of unprecedented scale, and the human and physical damage has been tremendous. Recovery and reconstruction have been slow, but there is mounting hope among victims of the earthquake and tsunami that a fresh start is on the horizon. However, in the case of Fukushima Prefecture, where the reactor meltdown and hydrogen explosions at the TEPCO Fukushima Daiichi nuclear power plant resulted in the radioactive contamination of air, land, and water, the situation is far more complex and daunting.

One of the most distinctive characteristics of a nuclear disaster is the nature of the evacuation that follows. While evacuation after an earthquake or tsunami is undertaken in response to devastation of homes and infrastructure, evacuation following a nuclear accident is undertaken to avoid potential devastation, namely exposure to radiation. Due to the speculative and anticipatory character of such evacuation, individuals and households have responded to the many agonizing questions it poses—including whether to evacuate, and if so, for how long and to where—in diverse and differentiated ways along the lines of gender, age, residential location, economic status, and so on. Moreover, since nuclear policies in Japan have long presumed that severe accidents were preventable and improbable, few municipalities had evacuation plans ready to respond to a nuclear accident of this magnitude.

One of the areas most severely afflicted by this nuclear accident was Futaba District, the remote coastal area where Fukushima Daiichi was constructed in the early 1970s. Futaba District consists of eight municipalities, including six towns (*machi*) and two villages (*mura*), many of which still remain classified as restricted areas due to high radioactive contamination. Due to the absence of systematic plans in these towns and villages, residents had to make their own evacuation decisions. This resulted in dispersed and discrete evacuation patterns in which individuals evacuated to distant locations and many family members continue to live apart from one another. The spontaneous nature of this evacuation has led to a troubling lack of accurate and reliable information about evacuees, greatly hindering planning and policymaking.

Furthermore, a number of problems emerged due to evacuation orders by the national government and their repeated revisions. In particular, strong divisions within communities have arisen based on the spatial boundaries of evacuation orders (Yamakawa 2013). On one side are areas where the conditions for residents to return have been met. In these areas, efforts to help returning evacuees make a fresh start are now well advanced. On the other side are areas to which residents cannot yet return. Evacuees from these areas have two options. The first is to leave their home communities behind by opting to permanently relocate. The second is to bide their time until evacuation orders are lifted while living in makeshift communities, variously called "temporary towns" or "second towns" (Funabashi 2014). The complexity of evacuation-zone designations and the uncertainty of future options for evacuees make it highly difficult for them to obtain the accurate and reliable knowledge needed to restore their livelihoods.

The Futaba Survey

At the outset of the nuclear disaster, the national government ordered the population living within 20 km of the Fukushima Daiichi Nuclear Power Plant, which covered large portions of the eight towns and villages of Futaba District, to evacuate.[1] It soon became apparent that this evacuation order would remain in effect for a prolonged period. In an attempt to ameliorate the situation and better understand the conditions of the evacuees of the Fukushima nuclear disaster, the Disaster Reconstruction Research Center at Fukushima University conducted a large-scale survey (hereafter the "Futaba Survey") of evacuees from the eight municipalities of Futaba District between September and October 2011.

As of September 2011, the number of citizens from Fukushima Prefecture who were evacuated as a result of the Great East Japan Earthquake and nuclear disaster totaled 87,686, according to the Fukushima prefectural government. The largest number of refugees was from Futaba District, 48,498 (55.3 percent) in total, followed by Soma District, just north of Futaba, at 24,715 (28.2 percent).[2] Clearly Futaba was the district most severely afflicted by the nuclear disaster. With cooperation from each local government in the district, questionnaire surveys were sent to 28,184 households by post, and completed surveys were received from 13,576 households (48.2 percent response rate). Responses were received from household heads and household members ranging from 17 to 103 years old. The survey was the first of its kind, with questions focused on the kind and extent of damage to property, living conditions at evacuation sites, expectations for return, plans for future living and livelihood, and opinions and concerns about the recovery and reconstruction of the afflicted areas and administrative support for evacuees. The results of the survey were published in the "Basic Summary of the Survey of the Reconstruction of the Eight Towns and Villages of Futaba in 2011 Vol. 2." This chapter provides a brief summary of the survey findings.[3]

Between April and August 2012, the evacuation-designated areas of the eight towns and villages of Futaba District were reorganized, based on annual cumulative radiation dose, into "difficult to return areas" (highly radiation-contaminated areas to which residents will not be able to return for a long time), “limited residence areas” (areas which residents may visit but in which they are not permitted to live), and “evacuation lift preparation areas” (areas to which residents will be able to return in the near future). Therefore, some areas of Futaba District are now witnessing the return of residents, and reconstruction is underway, however slow and in some cases problematic it might be. Consequently, the survey results should not be taken as describing the most up-to-date conditions of evacuees’ lives. Nevertheless, this snapshot of evacuees at the peak of mass evacuation offers us important insights into the nature of this historic disaster.

Overview of the evacuation process

The geography of evacuation

Unlike organized and group-based evacuations following natural disasters, the evacuation following the Fukushima nuclear accident was more discrete, more dispersed, and, in many ways, more chaotic. It is suggestive of the chaotic nature of evacuation that by the time of the survey, within seven months after the disaster, 47.2 percent of respondents had changed their evacuation location three or four times and 35.6 percent more than five times. With this in mind, let us first look at the geographic patterns of evacuation at the time of the survey and the reasons evacuees chose their destinations.

The Futaba Survey examined the number of evacuees from Futaba District received by each prefecture in Japan (Figure 4.1). In descending order, from most evacuees to fewest, these prefectures were as follows: Fukushima (69 percent), Saitama (6 percent), Tokyo (5 percent), Ibaraki, Chiba, and Kanagawa (3 percent), and Niigata, Miyagi, and Tochigi (2 percent). Excluding Tottori Prefecture, evacuees were living in every prefecture in the nation from Okinawa to Hokkaido. Looking at the figures for evacuees who relocated outside Fukushima Prefecture, we see that the percentages vary according to age, including 39 percent of individuals aged 20–39, 29 percent of individuals aged 40–9, 26 percent of individuals aged 50–9, and 32 percent of individuals aged over 60. In general, families with elementary-school-age children left the prefecture and moved further away from Fukushima.

As for their reasons for choosing where to move, the survey results indicate that 30 percent of respondents, the largest number, selected their destination because it was “near relatives and friends.” This reason was followed, in descending order, by “radiation concerns” (23 percent), “workplace considerations” (22 percent), “administrative guidance”[4] (16 percent), “lower economic burden” (16 percent), and “child and school considerations” (15 percent). Most importantly, this finding reveals the diversity of reasons for choosing particular destinations, including social, economic, and health concerns. Rather surprisingly, only 8 percent of respondents indicated that they chose their evacuation destination due to a desire to “be with former neighbors,” a finding that calls into question the recent emphasis of Japanese disaster policies on mutual support among local community

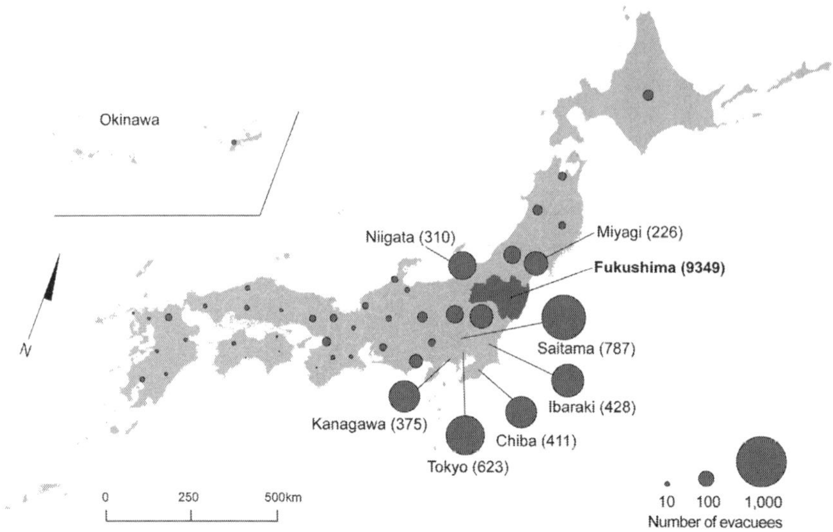

Figure 4.1 Distribution of evacuees by prefecture as of September–October 2011.

Source: Futaba Survey.

members (Yoshihara 2013). These findings demand that we critically reflect on existing policies for assisting evacuees.

The relative importance of reasons for choosing particular destinations varies by gender and age group. For example, women and individuals aged 40–9 tended to cite "children and educational considerations" as their most important reason, suggesting a strong urge to minimize the effects of radiation on children (see also Chapter 5), while other groups (men and individuals under 30 or over 50) tended to prioritize "workplace considerations." These differences in reasoning across gender and age groups are clearly factors underlying the discrete and dispersed evacuation patterns observed following the nuclear disaster.

Reasons for not returning

Perhaps one of the most striking, but also understandable, features of the nuclear disaster is that it produced a large number of evacuees who are not willing to return home. Of the 13,576 respondents, 3,284 (24 percent) answered that they were not willing to return home. Over 83 percent of these evacuees expressed a "disbelief in effectiveness of decontamination," far and away the most frequently indicated reason for not returning (Table 4.1). This was followed, in descending order, by 66 percent of evacuees who indicated that they did not want to return due to "distrust of government safety declarations," 61 percent who indicated "weak expectations for nuclear-accident resolution," and 42 percent who indicated "concerns over daily life and finance." Other reasons include "opposition

Table 4.1 Reasons for not wanting to return (percent, multiple answers allowed)

	Disbelief in decontamination	*Distrust of government safety declarations*	*Low expectations for accident resolution*	*Concerns over daily life and finance*	*Opposition from other family members*	*Found new employment*	*Total number of respondents*
Male	83.6	64.9	60.8	40.7	14.8	7.1	1,922
Female	82.1	66.6	61.7	42.8	14.2	7.2	1,333
10–19	60.0	60.0	60.0	20.0	30.0	20.0	10
20–29	78.8	68.6	60.8	39.6	16.7	12.7	245
30–39	84.6	68.3	57.1	38.0	13.7	12.4	597
40–49	84.9	71.3	58.8	42.7	12.3	9.5	536
50–59	85.2	64.9	61.5	47.9	13.0	6.9	792
60+	80.5	61.5	64.6	39.0	16.5	1.8	1,085
Total	82.8	65.6	61.2	41.5	14.5	7.1	3,284

Source: Futaba Survey.

from other family members" (15 percent) and "found new employment" (7 percent). It is noteworthy that the most important reasons do not arise from objective conditions of their home region (such as the level of recovery) but rather from a distrust of authority and official information.

The survey also asked evacuees to indicate the conditions that would have to be met before they would be willing to return to their former residences. It is noteworthy that a mere 4 percent of respondents indicated they would return "if radiation levels decrease to the level of the government's safety declaration," implying not only concerns about radiation but also distrust of the government's standards and policy stance. To the same question, 16 percent of respondents indicated "after reconstruction of infrastructure" and 21 percent indicated "after implementation of decontamination," both of which may be considered physical environmental factors. Interestingly, a sizable number of evacuees (26 percent of respondents) would return only if and when other Futaba evacuees returned, implying the importance of a social environmental factor. This answer may also indicate that evacuees would rather trust their neighbors than the government and experts in their judgment to return. Rather than interpreting this as a lack of rational, independent thinking, we must accept that an important part of the decision to return is socially motivated, and that what is needed is not simply a call for return based on scientific and objective safety standards but, no less critically, the establishment of a sense of confidence and reassurance even when scientific knowledge is uncertain or lacking.

We must also note that 25 percent of respondents indicated "no desire to return," and that there is a great age-based difference on this point. While 46 percent of those below 34 years old indicated "no desire to return," only 16 percent of those individuals aged 65–79 indicated the same. The significant proportion of evacuees, especially in younger age groups, unwilling to return regardless of the state of recovery (i.e. decontamination) needs to be taken into consideration in future reconstruction policies.

Obstacles and challenges confronting evacuees

Current and future obstacles

Evacuees have encountered various hardships, obstacles, and concerns at their evacuation destinations. The Futaba Survey examined how evacuees thought these concerns would change in the years ahead, by comparing and contrasting responses provided by evacuees to questions about "current" and "future obstacles." Asked to identify "current obstacles," 57 percent of respondents indicated "concerns over the effects of radioactivity" (Table 4.2). This, the most frequent response, was followed by, in descending order, "insufficient finances for daily expenditures" (34 percent), "lack of prospects for housing" (33 percent), "lack of work and employment" (30 percent), "declining health and supportive care" (26 percent), "relations with others" (25 percent), "worsening family relations" (18 percent), and "children's education" (15 percent). Perhaps not surprisingly, some of the answers show a clear gender-based difference. In particular, more

Table 4.2 Current obstacles and challenges (percent, multiple answers allowed)

	Concerns over radioactivity	*Insufficient finances for daily expenditures*	*Lack of prospects for housing*	*Lack of work and employment*	*Declining health and supportive care*	*Relations with others*	*Worsening family relations*	*Children's education*	*None*	*Total number of respondents*
Male	57.2	36.2	33.2	30.4	26.7	23.4	18.2	14.5	6.2	8,711
Female	55.2	29.2	31.5	28.6	25.3	26.6	18.1	17.2	5.4	4,758
10–19	48.6	11.4	11.4	14.3	5.7	8.6	2.9	20.0	20.0	35
20–29	62.8	36.3	24.2	27.8	15.0	33.1	18.4	22.9	6.8	532
30–39	61.6	31.7	32.2	27.5	16.9	31.5	20.3	39.8	3.3	1,456
40–49	55.0	37.3	32.6	35.8	21.6	27.0	22.3	37.4	4.0	1,860
50–59	54.5	35.1	31.8	38.1	25.9	22.7	20.4	11.8	4.9	3,192
60+	56.3	32.6	33.9	24.7	30.7	22.7	15.3	4.7	7.3	6,485
Total	56.5	33.8	32.6	29.7	26.2	24.6	18.1	15.4	5.9	13,630

Source: Futaba Survey.

women than men answered “relations with others” and “children’s education” were important obstacles and challenges.

When asked to identify “future obstacles,” 56 percent of evacuees described themselves as “unable to make decisions due to lack of knowledge about length of evacuation” (Table 4.3), the response most often given, indicating that uncertainty and the lack of predictability were leading concerns. Other important issues included “no plans for where to live in the future” (48 percent), “concerns about effects from radioactivity” (46 percent), and “lack of prospects for finances for daily expenditures” (30 percent).

What the above results clearly indicate is that the main issues that evacuees are dealing with shift as they transfer from evacuation centers to temporary housing. Their most pressing daily issues shift from concerns about radiation exposure to the long-term challenge of where and how to find adequate housing. Additionally, because initial criteria for emergency housing-placement decisions prioritized “individual” attributes such as age and vulnerability over “community” attributes such as family, friends, and neighborhood relations, at some sites evacuee populations are composed overwhelmingly of elderly individuals who have moved from distant locales, and many evacuees experienced great difficulty in forming interpersonal relations with fellow evacuees. This issue has been partially resolved over time through the improvement of volunteer support from not-for-profit organizations (NPOs) and consultation services from the government, which have led to the formation of new communities and helped to mitigate the problem. Nevertheless, serious issues—including the continued occurrence of earthquake-related deaths among the elderly and the critical issue of how to secure adequate and appropriate education opportunities for children—have not been solved (Chapter 10). Minimizing the potential for radiation exposure for populations in these areas is an obvious and essential prerequisite. However, without adequate support for daily life and employment, as well as health and education for children, it will be impossible to stem the outflow of populations from the evacuated areas.

As suggested by these data, nuclear-disaster evacuees’ main current concerns are related to issues of radiation, while their future concerns all relate to uncertainties about the period of evacuation, which is to say that their concerns are centered on the many uncertainties in their lives. Indeed, it is in the very nature of this nuclear disaster that no clear end is in sight. Rather than resorting to simplistic “plans for recovery,” policies to assist the livelihoods of evacuees must be designed on the premise of significant uncertainty.

On the difficulties of making life plans

Nuclear-disaster evacuees face challenges in dealing with daily expenses and finances. The Futaba Survey found that evacuees were relying on donations and provisional compensation for 81 percent of their expenditures, making this source their top means of income (Table 4.4). Donations and compensation were followed, in descending order, by a 39 percent reliance rate on pensions, a 34 percent

Table 4.3 Expected future obstacles and challenges (percent, up to three answers)

	Unable to make decisions due to lack of knowledge about length of evacuation	*No plans for where to live in the future*	*Concerns about effects from radioactivity*	*Lack of prospects for finances for daily expenditures*	*Concerns about ability to rely on relations with acquaintances and friends from former communities*	*Concerns about children's education*	*Inability to find work at evacuation destinations*	*Concerns about getting along well with others*	*Lack of prospects for doing business*	*Total number of respondents*
Male	57.4	47.5	45.6	31.8	18.9	13.2	13.8	11.9	11.3	8,711
Female	54.2	49.0	47.0	25.6	20.4	17.3	14.5	14.9	5.8	4,758
10–19	17.1	17.1	62.9	14.3	11.4	22.9	11.4	–	2.9	35
20–29	38.5	37.0	52.6	25.8	14.3	24.8	18.2	16.2	4.7	532
30–39	48.3	46.5	50.9	22.9	12.0	37.6	15.8	13.7	6.2	1,456
40–49	53.9	46.9	43.5	26.8	10.2	36.6	18.9	9.1	9.8	1,860
50–59	57.8	50.8	41.6	31.7	16.2	10.5	19.8	11.5	11.2	3,192
60+	59.7	48.4	47.4	31.4	25.9	4.2	9.0	14.4	9.5	6,485
Total	56.2	48.0	46.1	29.7	19.4	14.6	14.0	12.9	9.4	13,630

Source: Futaba Survey.

Table 4.4 How evacuees are making ends meet (percent, multiple answers allowed)

	Donations and provisional compensation	*Earned income*	*Revenue*	*Pensions*	*Savings*	*Debt/loans*	*Welfare*	*Total number of respondents*
Male	80.4	35.4	2.2	40.1	34.7	2.5	0.5	8,711
Female	81.5	33.9	1.2	37.3	33.1	1.1	0.9	4,758
10–19	45.7	17.1	–	–	20.0	5.7	2.9	35
20–29	81.4	64.8	1.1	2.8	38.5	1.1	1.3	532
30–39	81.0	66.8	1.6	2.6	36.1	2.3	0.7	1,456
40–49	81.2	61.8	1.9	6.7	35.9	2.3	0.1	1,860
50–59	79.4	52.0	3.0	9.1	34.1	2.8	0.5	3,192
60+	81.4	8.9	1.4	74.8	32.8	1.6	0.9	6,485
Total	80.7	34.7	1.8	39.2	34.0	2.0	0.7	13,630

reliance rate on earned income, and a 34 percent reliance rate on personal savings. Differences in the financial resources that evacuees are drawing on to fund their daily life expenditures can be found when we look at these figures by age bracket. In all age brackets, high rates of reliance on donations and provisional compensation payments as well as a devastating drawdown of personal savings can be seen. However, for evacuees aged 20–59, these sources are supplemented by earned income and personal savings, while for evacuees over 60 years of age they are supplemented by pensions. If we look at the data on financial resources according to residential type (not included in the table), what stands out is that evacuees living in "evacuation centers," "temporary housing," and "with relatives and friends" reported that 40–59 percent of their finances are from pensions, a figure which reflects the large number of elderly living in these residential types. Additionally, for evacuees living in "government subsidized rental units" and "non-subsidized rental units," the high rate of reliance on earned income stands out, reflecting the fact that individuals living in these residential types are primarily between the ages of 20–59.

Examination of employment figures before and after the earthquake further illustrate the extremely difficult financial situation that evacuees are facing. The overall unemployment rate nearly doubled after the accident, from 28 percent before the disaster to 54 percent after. It is likely that nearly all unemployment before the accident was due to retirement, and that the increased figures were due to loss of employment as a result of the nuclear disaster. Indeed, while 33 percent of employees were full-time employees before the accident, this figure dropped to 20 percent after the accident. Additionally, self-employed workers dropped from 15 percent to 4 percent, part-time workers from 9 percent to 4 percent, government workers from 4.4 percent to 3.6 percent, and organizational staff from 2.1 percent to 1.4 percent. As this suggests, full-time employees, the self-employed, and part-time workers in the private sector—where the largest number of people were employed—have witnessed particularly large increases in unemployment. In terms of gender, unemployment for males increased from 24 percent to 49 percent and for females from 36 percent to 65 percent, a particularly drastic increase in unemployment for women. Indeed, for females, full-time employment and employment as organizational staff decreased by 50 percent, while self-employment and part-time work decreased by 25 percent.

On the health conditions of evacuees

The increasing severity of the health issues confronting evacuees is represented, in brief and concentrated form, by what are known as "earthquake-disaster-related deaths."[5] Three and a half years after the nuclear disaster, the number of earthquake-disaster-related deaths in Fukushima Prefecture reached 1,793. Looking at these figures in relation to age, we see that in each of the three prefectures the elderly represented a high percentage of all such deaths: 87 percent in Iwate Prefecture, 87 percent in Miyagi Prefecture, and 91 percent in Fukushima Prefecture. The timing of these earthquake-disaster-related deaths showed slight but important differences. In Miyagi Prefecture most earthquake-disaster-related deaths occurred in the first 7–30 days and in Iwate Prefecture in the first 30–60 days, but in Fukushima Prefecture most occurred between sixth months and a year after the earthquake (Reconstruction Agency 2014).

A study of 734 of these deaths in Fukushima determined that around 30 percent were caused by "physical and psychological exhaustion experienced during evacuation life" and around 20 percent by "delays in initial treatment due to hospital shutdowns." Furthermore, a summary of findings from individual case studies found that most of these deaths involved gradual weakening due to the initial evacuation after the nuclear disaster and the later stress, lethargy, and medical circumstances stemming from evacuation life. As medical personnel and public health officials expressed it,

> The greatest difference between the effects of the disaster in Iwate and Miyagi Prefectures and in Fukushima Prefecture is that, in Fukushima, it can be seen that the long-term difficulties of evacuation life lead people to agonize over whether they "may never live to see beyond evacuation center life" and they are left without purpose, hope or the will to live, in other words seriously troubling mental effects.
>
> (Reconstruction Agency 2013)

Prospects for returning home

While there are evacuees who do not to return to their former homes and villages, there are also many evacuees who do. When asked to identify their reasons for wanting to return home, 62 percent of evacuees indicated "affection for hometown" (Table 4.5). This sense of affection for home was followed, in descending order, by 58 percent of evacuees who cited "ancestral land, house, and graves," 40 percent the wish "to pursue reconstruction alongside my neighbors," 38 percent

Table 4.5 Reasons for hoping to return (percent, multiple answers allowed)

	Affection for hometown	*Ancestral land, house, and graves*	*To pursue reconstruction alongside my neighbors*	*Fond of living in the area*	*Concerns about moving to an unknown place*	*No desire to move to another location*	*Because family and other community members want to return*	*Total number of respondents*
Male	63.0	58.9	42.5	38.6	32.7	30.1	9.0	6,570
Female	61.2	55.6	36.3	37.1	37.2	28.0	9.9	3,301
10–19	36.4	4.5	36.4	13.6	9.1	9.1	–	22
20–29	58.1	30.6	26.1	39.1	29.9	16.2	7.7	284
30–39	63.1	41.8	29.4	39.5	32.4	19.8	11.5	837
40–49	60.8	54.9	34.3	34.2	30.2	30.6	10.6	1,293
50–59	60.0	58.6	39.0	34.2	32.2	30.7	8.0	2,332
60+	63.9	62.7	45.3	40.7	36.6	30.8	9.5	5,180
Total	62.3	57.9	40.4	38.1	34.1	29.3	9.4	9,992

Source: Futaba Survey.

"fond of living in the area," 34 percent "concerns about moving to an unknown place," and 29 percent "no desire to move to another location." It is possible to suggest that nearly all the reasons noted by respondents can be characterized as a sense of "local identity." Looking at these responses according to age, we find that the total number of reasons generally increased along with increases in age. This suggests that advances in age are correlated with an increasingly strong sense of local identity.

How long are evacuees willing to wait to return home? When they were asked that question in October 2011, 12 percent of respondents indicated "less than one year," followed by "one to two years" (35 percent), "two to three years" (22 percent), "three to five years" (11 percent), and "as long as it takes" (14 percent) (Table 4.6). Looking at the responses according to the towns from which evacuees originated, we find that those indicating "less than one year" were relatively numerous among evacuees from towns dealing with relatively low radiation levels, while the number of respondents indicating "three to five years" was relatively high for Futaba Town, Okuma Town, Tomioka Town, and Namie Town, all close to the nuclear plant.

The survey asked those evacuees who were unable to return to their home in the near future where they would live for the time being. The largest number of respondents indicated that they would relocate to the towns and villages bordering Futaba District (40 percent), while 19 percent indicated that they would relocate to towns and villages in Naka-Dori and Aizu, regions within Fukushima Prefecture but not bordering Futaba District. In total, 10 percent of respondents indicated that they would move outside Fukushima Prefecture, while 7 percent indicated that they would stay in Futaba District, and 16 percent were still undecided (the remaining 8 percent provided various other answers).

Needless to say, whether evacuees can return to their homes within their hoped-for timeframes depends heavily on the progress of decontamination and other reconstruction work. Starting in April 2012, original evacuation areas in the cities

Table 4.6 Duration willing to wait to return to former communities (percent)

	Less than one year	*1–2 years*	*2–3 years*	*3–5 years*	*As long as it takes*	*Unsure*	*Total number of respondents*
Male	11.9	36.1	22.4	10.5	13.0	6.2	6,570
Female	11.9	32.5	21.6	10.6	14.6	8.7	3,301
10–19	36.4	18.2	0.0	13.6	13.6	18.2	22
20–29	8.8	19.7	25.0	10.6	24.6	11.3	284
30–39	8.1	28.7	22.1	10.2	21.1	9.8	837
40–49	9.7	33.7	19.2	12.4	16.9	8.0	1,293
50–59	9.9	34.3	22.5	12.8	13.2	7.4	2,332
60+	14.1	37.2	22.7	9.2	11.0	5.8	5,180
Total	12.0	34.8	22.1	10.5	13.5	7.0	9,992

Source: Futaba Survey.

and towns in and around Futaba District have been reorganized, and parts of those areas are being prepared to eventually allow residents to return. Yet a number of problems remain unresolved on the ground, including how to differentiate financial compensation based on area designations when levels of radioactive contamination may vary greatly at micro scales (Chapters 8 and 10).

Views on reconstruction

The fundamental difference between reconstruction agencies' and victims' perspectives on reconstruction is that the former emphasize "creative reconstruction" while the latter are seeking "revitalization of their former lives." The phrase "creative reconstruction" describes a complicated process that begins with good intentions but often ends with business as usual. Prior to the disaster most towns and villages had "dormant" town plans, visionary plans that were devised by planning agencies but have been proven hard to implement due to difficulties in reaching agreement with landowners and bottlenecks with funding. With a major societal crisis, such as the current disaster, various special reconstruction laws and administrative measures are instituted, such as the relaxation of land-use regulations and the injection of public funding into local areas. Soon reconstruction begins to be seen as a "golden opportunity" to push through the dormant plans, and this opportunity for profit becomes the central and defining purpose of reconstruction planning. Thus, in the end, reconstruction-oriented community planning becomes little more than business as usual, a business venture led by consultants, real-estate agents, and, in particular, construction companies (Noda 2017).

In contrast to this "creative reconstruction," let us consider here the type of community building that victims of the disaster are inclined to favor. In the survey, victims were asked to indicate what they feel is "important for reconstruction." The number one overall response was "comprehensive reconstruction planning for the entire Futaba district" (46 percent), followed by "promotion of industrial restoration to create employment for younger people" (43 percent), and "enhancement of elderly and medical facilities" (31 percent). In the current situation all of the local municipalities have formulated long-term "comprehensive plans." However, while a well coordinated, district-wide plan for reconstruction is essential, funding for implementing "regional plans" that would extend over multiple municipalities has been limited to certain operations, thus resulting in a patchwork of local municipal "comprehensive plans" with only limited regional coordination.

Nevertheless, caution is required when interpreting the meaning of "comprehensive" in evacuees' number one response. The term should not be taken to indicate respondents' desire for "drastic" restructuring of their lives; rather, it implies "holistic" or "well coordinated" planning actions. Moreover, this does not necessarily imply that the communities of Futaba should be merged into one municipality (akin to the national government's push for municipal mergers in the early 2000s), since only 12 percent of respondents indicated that "merger of Futaba District into one municipality" was essential to reconstruction.

In reality the intra-village community serves as the foundational basis of the daily lives of villagers, and for that reason village-level community plans can play an important role in promoting the autonomy of villagers and improving their living environments. In practice, however, with the exception of Iitate Village, the formulation of such plans has not progressed. The fact is that the evacuation of villagers and the hollowing out of the village have hindered recovery and reconstruction from the earthquake and the nuclear disaster. In this sense, village-level community planning, to make the community more livable and attractive, has become even more important than before the disaster.

Conclusion

Five years have now passed since the disaster began, and while decontamination and reconstruction planning are beginning to be pursued as business opportunities by interested parties, the revitalization of the daily life of the people and communities affected by the nuclear disaster has advanced at an agonizingly slow pace. This chapter has explored the conditions of and difficulties faced by evacuees, whose lives have been in prolonged suspension, focusing on those from Futaba District, the most severely afflicted area of the nuclear disaster. The Futaba Survey has revealed a number of important characteristics of the nuclear-disaster evacuation, including dispersed destinations, variegated housing types, distrust in authority, differences in views and opinions across age and gender groups, and the presence of both hope and despair for return.

What are the implications of the survey for reconstruction policies? First, given the financial difficulties that evacuees face, proper compensation is the most urgent matter. There has been much speculation among victims that finding a new job or restarting business operations before compensation payments are made will hinder their efforts to receive compensation for their losses. The lack of transparency of process has certainly been one factor that has made nuclear-disaster victims reluctant to rebuild their lives.

Second, many evacuees had little choice but to live in dispersed government-subsidized and non-subsidized private rental units, and their actual lives are starting to take root in these areas. Accordingly, government-led, large-scale projects to build permanent or temporary collective housing and brand new communities at this point will be inadequate for the needs of evacuees and victims. Rather, unlike orthodox reconstruction policies, it is imperative that policies for nuclear-disaster evacuees accept and accommodate the dispersed and fluid nature of evacuation. This may mean, for example, much smaller-scale construction of public housing in vacant plots of land within the urban areas where evacuees already live. Policy support to link the improvement of evacuees' lives with community-building measures at evacuation sites is also required.

Finally, it is important to expect national and local governments not only to provide direct funding to support reconstruction efforts but also to establish community fund systems that allow the pooling of one-time mass compensation payments in order to carefully manage and utilize the funds for the long-term wellbeing of evacuees and afflicted communities. This is one way to cultivate local autonomy and capacity for self-governance, a move away from past dependence on "nuclear money," while at the same time holding the government and TEPCO responsible for the losses caused by the accident.

Acknowledgments

This chapter is a summary of the support activities for nuclear-disaster reconstruction provided by the FURE support center at Fukushima University, where I served as first director from July 2011 to March 2013. I would like to thank the faculty of Fukushima University involved with the center, and in particular the special staff and researchers at the center. This study was supported by JSPS KAKENHI (Multi-year Fund) Grant Number 2522043 (Grant-in-Aid for Scientific Research (S), 2013–2017, *Establishment of Academic Framework of Earthquake Disaster Reconstruction Experiencing Great East Japan Earthquake*, principal investigator: Mitsuo Yamakawa).

Notes

1 The evacuation designations within Soma District, which includes four municipalities, were more variegated internally due to different observed levels of radioactive contamination. While there were no "restricted area" designations in the Iwaki region, damage from the tsunami and earthquake resulted in the evacuation of 5,971 people from the district. In the Naka-Dori region, parts of Tamura City, Kawamata Town, and Date City were designated as evacuation areas, and 4,902 people were evacuated. In the Aizu region, "restricted areas" were not designated and there were no evacuees.
2 The administrative district of Soma is composed of four towns and villages: Shinchi Town, Soma City, Minami-Soma City, and Iitate Village.
3 A summary of the results (in Japanese) is available at: http://goo.gl/ApGQgp.
4 Administrative guidance (*gyosei shido*) is a notion that frequently appears in the study of Japanese administrative laws. It takes the form of a government agency giving advice, suggestions, instructions, and warnings to individuals and businesses, and these "are without statutory basis and are frequently made behind closed doors without written records being kept" (Sugimoto 2003: 217).
5 The category of earthquake-disaster-related death refers to the fatality not as the direct result of the disaster (for example, the collapse of buildings) but as the result of becoming ill during evacuation or of worsening injuries and chronic disease in poor living conditions.

References

Funabashi, Harutoshi. 2014. "The Third Pass for Reconstruction from the Disaster and Reform of Constellation of Arenas." *Trends in the Sciences*, 19(4): 82–7. [In Japanese.]

Noda, Takehito. 2017. "Why Do Local Residents Continue to Use Potentially Contaminated Surface Water after the Nuclear Accident? A Case Study of Kawauchi Village, Fukushima." In *Rebuilding Fukushima*, edited by Mitsuo Yamakawa and Daisaku Yamamoto, London: Routledge.

Reconstruction Agency. 2013. "Fukushima Ken ni okeru Shinsai Kanrenshi Boshi no Tameno Kento Hokoku" [Report from a Study on How to Prevent Earthquake Related Deaths in Fukushima]. Accessed January 27, 2016. www.reconstruction.go.jp/topics/20130329kanrenshi.pdf. [In Japanese.]

Reconstruction Agency, Cabinet Office (Disaster Management) and Fire Department. 2014. "Higashi Nihon Daishinsai ni okeru Shinsai Kanrenshi no Shishasu" [The Number of Earthquake-related Deaths from the Great East Japan Earthquake (as of September 30, 2014)]. Accessed January 27, 2016. www.reconstruction.go.jp/topics/main-cat2/sub-cat2-6/20151225_kanrenshi.pdf. [In Japanese.]

Sugimoto, Yoshio. 2003. *An Introduction to Japanese Society*, Cambridge: Cambridge University Press.

Yamakawa, Mitsuo. 2013. *Gensaichi Fukko no Keizai Chirigaku* [The Economic Geography of Restoration after Disaster], Tokyo: Sakurai-shoten. [In Japanese.]

Yoshihara, Naoki. 2013. "Genpatsu Sama No Machi" *Karano Dakkyaku: Okuma Machi Kara Kangaeru Komyuniti No Mirai* [Departing from the "Nuclear Town": Thinking about the Future of Communities in Okuma Town], Tokyo: Iwanami Shoten. [In Japanese.]

5 Displacement and hope after adversity

Narratives of evacuees following the Fukushima nuclear accident

Naoko Horikawa

Please pass on our stories to the next generation and do not let our existence be a secret.

(Woman who voluntarily evacuated from Koriyama City to Tokyo)

I really think about human security after this nuclear accident. Why didn't the Japanese government give us the truth at the time about how we might be affected by radioactive contamination?

(Woman who voluntarily evacuated from Fukushima City to Tokyo)

Introduction

On March 11, 2011, an unprecedented disaster was etched into history as the Great East Japan Earthquake, the most powerful earthquake ever to hit the country and which triggered a massive tsunami that took the lives of thousands and devastated the land on which they lived. The tsunami also resulted in severe damage to the Fukushima Daiichi Nuclear Power Plant (NPP) operated by the Tokyo Electric Power Company (TEPCO). An estimated 150,000 residents of Fukushima Prefecture were displaced as the consequence of both real and perceived risks of radioactive contamination.

Despite intensive decontamination programs and the national government's drive for return, approximately 100,000 people remained in a state of evacuation three years after the disaster. As of March 2014 about 47,000, including 12,000 children, were believed to be living outside the prefecture. The largest number resided in Tokyo, 6,051 (13 per cent), followed by Saitama Prefecture 5,011 (11 per cent), Yamagata Prefecture 3,960 (8 per cent), and Ibaraki Prefecture 3,474 (7 per cent), although evacuees could be found in all prefectures of Japan (Fukushima Prefecture 2016). These evacuees now live with their friends and relatives, in various forms of public or private housing, or in temporary accommodation, including hospitals. Of the total, some 20,000 are considered "voluntary evacuees," implying that their translocation is a matter of personal or family choice.

Research on the Fukushima disaster has generated and continues to generate a large volume of studies, ranging from biophysical research on the effects of radioactive contamination and impact on public health to social-scientific exploration of the reconstruction process and the manner in which residents cope with their predicament. In the social-science literature, Yamakawa (2013) investigates the situation of affected regions from an economic geographic perspective, while Yamashita *et al.* (2013) focus on the concept of evacuation and evacuees. With respect to the issue of evacuee status, Imai (2014) suggests that evacuees be given a kind of dual citizenship. Seki and Hiromoto (2014) describe narratives of evacuees from Fukushima in Tosu City in Saga Prefecture, highlighting the pains of adjustment and feelings of ambivalence. Yamane (2013), using a questionnaire survey, describes the lives of voluntary evacuees in Yamagata Prefecture, their difficulties in raising a family, and their uncertainty about economic circumstances. Ikeda (2013), on the other hand, draws attention to how mothers in Fukushima and Koriyama City accept the risks of radiation and negotiate information on contamination.

One of the distinct forms of voluntary evacuation that gained media attention during the time of nuclear disaster was "mother-and-child evacuation" (*boshi hinan*). This typically means that mothers, fearing the potential effects of radiation on their children, move away from the family home while husbands and other household members such as parents-in-law either stay or go to different locations (see also Chapter 8). Because this form of evacuation is not widely seen in other disasters, there is little research on it, and post-disaster policies may not be catering to the needs and demands of these evacuees; hence it is important to gain a thorough understanding of their experiences and perceptions.

In this chapter I explore the experiences, thoughts and feelings, and coping strategies of voluntary evacuees, many of whom are "mother-and-child evacuees," in the wake of the Fukushima nuclear accident. By doing so I hope to shed light on the lives of ordinary people caught up in the nuclear disaster and its aftermath. Field research was conducted between October 2014 and May 2015. During the period I visited various events for evacuees such as support-group gatherings organized by volunteer supporters and evacuees themselves in different areas of Tokyo, Kanagawa, and Fukushima Prefectures. Through participatory observations and semi-structured interviews with 30 evacuees,[1] I explored how they coped with the event on March 11, 2011, their reasons for deciding to evacuate voluntarily, the processes and challenges during multiple relocations, the changes and adjustments in family relations, the adjustments required at evacuation destinations, and their future plans. I also asked their views on government policies and support and how these influenced their decisions and resettling processes. It was not always easy to gain information because many of the evacuees did not wish to let their experiences be known, and I made the utmost effort to build rapport with them first. Most interviews lasted about two hours and were audio recorded with the permission of the interviewees. Many mothers who voluntarily evacuated with children have experienced fear, uncertainty, and profound change in their everyday lives through dislocation. They also share a sense of distrust in authority and even in scientific knowledge. Beyond these generalities, their experiences vary widely as described below.

Forms of evacuation

Yamashita *et al.* (2013) distinguish three types of evacuation as the result of the Fukushima nuclear disaster: forced, voluntary, and "in-house." First, forced evacuation refers to the relocation of residents called for by the government following the hydrogen explosion at the Fukushima Daiichi NPP, which resulted in the diffusion of radioactive materials (Chapter 2). The initial evacuation zone, on March 11, 2011, included areas within 10 km of the Fukushima Daiichi NPP. The zone was extended to a 20 km radius on March 12 after the hydrogen explosion of another reactor building. Three days later, following additional explosions and a fire, residents living between 20 and 30 km from the plant were advised to stay indoors. Communities affected by the forced evacuation orders were: the towns of Okuma, Futaba, Tomioka, Namie, Naraha, Hirono and Kawamata; the villages of Kawauchi, Katsurao, and Iitate; Minami-Soma City; and parts of Tamura City. In these cases local government assisted evacuation by providing mass transportation as well as coordinating group-based relocation to evacuation shelters and temporary housing, although relocation processes were by no means smooth and orderly.

Second, voluntary evacuation is the relocation of people from their homes by processes not ordered or directly assisted by governments. Within a few days of the initial explosion, many residents living outside the officially declared evacuation zones in areas such as Fukushima City, Koriyama City, and Iwaki City decided to leave. The prevailing view is that these evacuees were mothers with children concerned about exposure to high levels of radiation, and that their husbands who had jobs in Fukushima would remain. This is an overly generalized view, and there are significant variations, including cases of a father who evacuated with children while a mother remained in Fukushima Prefecture, or both parents and children making the move together as a family. There are also cases of voluntary evacuation of single men.

The third type of evacuation, "in-house" evacuation, does not actually involve the physical relocation of people. This term refers to the residents of Fukushima, outside the mandatory evacuation zones, who practice various measures to protect themselves from possible effects of radiation by cautious selection and consumption of food and water, wearing masks outside, and sending their children away during school breaks. Some would not consider this a form of evacuation. Yamashita *et al.* (2013), however, include it as one because in essence these "evacuees" are distancing themselves from an environment in which they perceive high risks. Although the focus of this study is the second form of evacuation, voluntary evacuation, it is important to bear in mind that "in-house" evacuation is often the choice forced up on those who cannot voluntarily evacuate, as illustrated by one of the informants I encountered:

> We obtained a mortgage and built a new house just one year before the accident. Our daughter was only two years old at that time. However, we decided not to move. The area suffered relatively high contamination. I continually ask for our home to be decontaminated because the reading for the upstairs is 0.3 μSv per hour, but I think the mayor of our city ignores this area.

In similar fashion, a number of residents of areas exposed to levels of higher than normal radiation contemplated moving out, and would have done so had it been feasible, but felt they had to stay and cope with the risks on a day-to-day basis. For many, it would seem that "in-house" evacuation acted as a substitute for voluntary evacuation.

Stories of mothers: how they made the decision to evacuate voluntarily

I now describe stories of voluntary evacuation, identifying variations within the type and looking at the factors that underlie such variation. The approach is meant to complement surveys of a more systematic nature (Chapter 4).

Mother's instinct

Kazuko, in her late 30s, from Fukushima City, explained how from the outset of the accident she felt differences with her husband regarding the idea of evacuation. When she saw the media coverage of the explosion at the Fukushima Daiichi NPP, she thought immediately that it would be a disaster for the prefecture and decided to move away as soon as possible. Although she lived more than 50 km from the plant, she took the situation seriously and instructed her children to stay indoors.

> Our house was sited near the Ground Self-Defense Force base. The road outside became so busy and noisy, as if there was a war. I can still hear the sound of helicopters. It really stressed me out because I felt that the situation was not normal and was getting worse. So I suggested to my husband that we evacuate together immediately. My second daughter was just one year old and I knew it would be difficult staying in sheltered accommodation.

Clearly, Kazuko was immediately fearful of the risk of radioactive contamination and desired to remove her children from the area without delay. The family evacuated to a relative of her husband's in Tokyo, staying there for two weeks. She became keenly aware of the vast difference between Fukushima and Tokyo in the ways that people carried out their everyday lives. It was an impression she repeated to me—daily life in Tokyo contrasted sharply to the situation back home. Ten days later they rented an apartment and her husband returned to Fukushima. The husband thought that his wife and children could also return relatively soon, although more than three years later Kazuko and their children remain in Tokyo.

Such conflicting assessments of the risks are explained as a characteristic gender difference by Ikeda (2013: 165).

Surrounding emotions and faith in authority

Another case indicates how the decision to evacuate was made not only on the instinct of a mother but was shaped by the surroundings. Kashimura (2011), taking a Lacanian approach, points out that in a situation of panic or urgent predicament many people tend to be influenced by others emotionally rather than logically. Yuka, in her early 40s, from Iwaki City, is a single mother with two daughters. She told me that her friends urged her to evacuate from Iwaki City as soon as possible. One close friend insisted, "Hurry up, hurry up." Many had already left. Meanwhile, her parents were telling her that the government was saying that the situation was not urgent and everyone would be fine. However, it began to worry even the parents that information given out by the authorities was mixed and confusing, and they began to doubt whether they were receiving the right advice (Chapter 3). Eventually her parents thought it might be a serious matter after all, so they all decided to evacuate from Iwaki. This was three days after the explosions at Unit 2 of the nuclear plant.

> We took a taxi to Tokyo, planning to stay at my uncle's house. However, my uncle refused to let us stay with him so we had to book a hotel for the night. We changed accommodation four times: hotels and sheltered accommodation. Finally, after six months, we moved into a collective house.

Yuka's case illustrates the emotive dimension of decision-making and also the shift (i.e. deterioration) in trust in the authorities over information regarding the situation.

Such deteriorating confidence in authority was directed not only towards the national government and experts in the mass media: it was seen locally. Kumi's house was a log cabin in Otama Village, some 60 km from the Fukushima Daiichi NPP. Kumi told me that she had no clear memory of March 11, 2011, but she remembers March 14, when there was a hydrogen explosion at Unit 3 of the plant. She had been queuing at a gas station for two hours, had a sore face, and was coughing awfully. That day, she said, was a nightmare.

> I was not aware of the level of radiation at the time. On 18 March we went to my parents' house in Sagamihara City, Kanagawa Prefecture. My five-year-old daughter and I stayed there for one month, returning once to Otama Village to attend an enrollment ceremony at a kindergarten. There was no specific information from the director of the kindergarten regarding radioactive contamination. His reply to the question from a parent [about any precautions] was that the children should stay indoors and wear a mask. I was so concerned at hearing this and took my daughter back to my parents' house.

Before the disaster, Kumi had given no thought to how a nuclear plant worked, nor to the risk of radiation. However, the sober words of the director of the kindergarten somehow seemed absurd and uncaring, and Kumi could not simply accept them at face value. Kumi and her husband discussed their daughter's education and her wellbeing and decided that it would be best for her and her mother to leave Otama. Their daughter was willing to go to wherever her mother would be, and that made their decision relatively easy.

A decision delayed

Junko, in her early 40s, made a decision a year after the start of the nuclear disaster. In April 2011 Junko had just started a new job near her home, which she very much wanted to keep, and therefore she did not wish to consider evacuation. She said that she had little knowledge of radioactive contamination at the time. A year later, after assessing a variety of information on the internet, she came to think that the local environment was not safe for children. A growing anxiety drove her to the conclusion that it was time to leave Fukushima. She found that there were some collective houses for evacuees in Tokyo, near areas where she had relatives, and she often visited them. Her husband remained in Fukushima because his age meant it might be difficult for him to get a suitable job in Tokyo. This meant that they must live apart and make corresponding financial adjustments. Junko moved into a municipal housing unit with two of her children, and her elder daughter joined them one year later after having finished secondary school in Fukushima. This case illustrates how voluntary evacuation might not have been an instantaneous reaction; rather, for some, it was an outcome of careful, though extremely agonizing, consideration.

Who moves and who stays?

In a typical example of voluntary evacuation, Mr Yoshi stayed while his wife and daughter moved to Kyoto. Mr Yoshi told me that he has a job in Fukushima as a member of a non-profit organization that oversees a recuperative program for children and other voluntary work, such as supporting evacuees.

> I guess our case is not exceptional, where a father elects to live alone for his job. The situation derived from the nuclear explosion, and we were forced to choose for the safety of my wife and daughter. I really like Fukushima, in which I was born and brought up until leaving for Tokyo for university, where I met my wife. I returned to Fukushima because I am an elder son. Now I live with my aged mother. Actually, my feelings are very complex. My wife and daughter do not share the same sentiment towards Fukushima as a hometown. They can live wherever there is an assurance of safety and an intuitive appeal for them. In a sense they are resilient and strong. As for myself, I think that Fukushima is the best place for me.

Mr Yoshi's narrative is perhaps representative of the image that many of us have of voluntary evacuation. Nevertheless, this is by no means the only form it takes.

In contrast to Mr Yoshi's case, what follows may seem unusual, in that a mother remained in Fukushima while her husband and their two daughters evacuated. Akiyo, in her late 40s, and her husband are both pharmacologists. She had been working at a pharmacy, located in an area of above-normal levels of radiation. Her husband and daughters moved to her parents' home in Hokkaido, and their decision to evacuate was primarily based on their concern over their girls' health and education.

> My family had to change lifestyle because of this accident. All the same, I do not consider the situation solely in a negative light. Our final decision was made one year later. We discussed so many times how to protect our children's health from contamination by radiation and how to provide them with a good education. I have been learning about radioactive contamination, reading books, going to seminars, and watching films concerning the Fukushima Daiichi Nuclear Power Plant. Ideally, children should be brought up by both parents, but in this predicament we chose an alternative lifestyle.

Akiyo's house was built several years ago on a site belonging to her husband's parents in Fukushima City. They have a mortgage to pay off. In the end, they decided that her husband would take a job in Hokkaido and she would remain in Fukushima. She decided to stay because she wanted to have direct knowledge about radioactive contamination and also felt that she had a useful role in her community as a pharmacologist. Their children's attitude also helped them make the decision:

> My children wanted to stay at my parents' house, so I felt relieved when they moved to Hokkaido. I would have carried a sense of guilt had we kept the children in Fukushima for another year. I will see them all regularly until my youngest daughter leaves high school.

Akiyo believes that this arrangement may actually work better than the typical arrangement, whereby the mother and children leave and the husband stays in Fukushima. The typical arrangement often leads to a perceptual gap between the husband (who considers that Fukushima is safe once more) and the wife (who wants to stay away from Fukushima unless there is absolutely no doubt that levels of radiation are safe), a situation which may subsequently result in the wife and children only reluctantly returning to Fukushima after a year or two. Akiyo did not have to face that problem. She said later on, "I am sure that our decision is right for our family, and I came to think that I am able to endure the present family life."

Family reunited and family disintegrated

As time has passed since the beginning of the accident, the household arrangements of some families have undergone changes. For example, Akemi, in her mid 30s, initially evacuated from Koriyama City to Sado Island, Niigata Prefecture, together with her two sons. She made the decision to move out quickly. Her husband seemed too engaged in his work.

> I intended to stay only for the summer of 2011. But the situation in Fukushima did not improve much over the summer, and decontamination of the area around our home was not complete.

They remained on the island for two years. During that time Akemi thought about the situation, concluding that the family should live together, and that her sons needed their father. Perhaps the period of separation presented a watershed for considering the family's future.

> I discussed the situation with my husband several times, and whether or not we should continue the arrangement. My fear was that if we carried on in the present way then our family would disintegrate.

Fortunately, her husband secured a job in Tokyo, and Akemi moved there also, enabling the family to be reunited. Recently they acquired a house of their own. Akemi continues to reflect on whether her decision to evacuate was the right one. She told me that the worst is when her elder son cries, saying that he really wants to go back to Fukushima, or to Sado, and not be in Tokyo.

Sometimes the disruption resulted in a marital breakdown. In the case of Tomo, in her late 30s, for example, her husband felt that one of the conditions of the marriage should be to look after his aging parents. He suggested Tomo should evacuate temporarily to Tokyo because she was pregnant at the time, while he remained in Date City, Fukushima. Tomo shuttled from Date City to Tokyo some ten times during her pregnancy. The arrangement did not last and they ended up divorcing.

> My parents were strongly against our divorce, saying that you should come back to Fukushima, as every member in our family lives in Fukushima. They think that I should obey them and my husband's opinion. I believe that this is typical parochial thinking. I always had to consult my husband about the continuation of the evacuation. My husband encouraged me to evacuate from Fukushima but wanted me to come back to Fukushima after the delivery of the baby. I really thought that I had to protect my child from radioactive contamination. Nobody taught me how to protect my baby and myself because nobody had experienced such a disaster. It took a year and a half to get divorced. I had panic attacks several times during the process.

Tomo added that her life in Tokyo as a single mother was better than in Fukushima because of the freedom she has.

Lives in voluntary evacuation

Living in limbo

When contemplating the future my interviewees see three options available. These are: remain in a present place of evacuation, move to another place, or return to Fukushima Prefecture. As far as voluntary evacuees are concerned, the decision is always in their hands. Imai (2014) refers to the situation as neither temporary nor permanent and suggests that until a decision is made, which in some instances may take as long as ten years or more, the government should provide sufficient support. However, the cost of living is a crucial factor. Tokyo is an expensive place to live. The decision thus depends partly on the availability of housing benefits. In the case of Tokyo, the municipal authorities provide free housing for both voluntary and forced evacuees, who also receive a monthly cash allowance, with a time limit now extended from March 2016 to March 2017. All my interviewees oppose the ending of the program. Some are applying for municipal housing in Tokyo and are considering settling there, with children's education an important factor. Others plan in time to return to Fukushima, where there are signs of improvement. One interviewee said, however, that no matter how much decontamination work is conducted she does not believe Fukushima will ever be safe for children. There are those who have yet to decide on whether to stay or return.

For many mothers with children who relocated as a result of the Fukushima disaster, life in limbo is not easy. Kazuko (mentioned above) told me, "We are poverty stricken because of the dual life between Fukushima and Tokyo." She speaks openly about the experience of evacuees. When she attends events such as symposiums she sells small articles like handkerchiefs. Some evacuees apply for assistance through Alternative Dispute Resolution (ADR), but Kazuko is not interested in mediation. Rather, she aims to be financially independent and seeks support for vocational approaches and allowances for voluntary evacuees.

The issues of compensation

There is a considerable gap between the levels of compensation for forced and voluntary evacuees (see also Chapter 10). Evacuees are regarded as "forced" if their original residence is located within evacuation zones designated by the government. This means that once the designation is removed, after decontamination work lowers radiation below specified levels, those who were previously considered forced evacuees suddenly become voluntary evacuees, incurring a drastic drop in financial compensation. In reality, many evacuees are unwilling to go back immediately due to lingering concern about low-level radiation, lack of essential infrastructure (for example, hospitals, shops, and schools), and the absence of neighbors and friends (Chapter 4). The status of one of the evacuees I met has changed from 'forced' to 'voluntary,' and she no longer receives the monthly payout of 100,000 yen, although she can still cover her apartment rent with the 'voluntary' compensation policy. Accordingly, she has decided to stay in Tokyo until she can foresee a normal life back home.

The current situation raises the question of how evacuees should be compensated. Yoshimura (2015) argues that voluntary evacuation should be treated in the same manner as forced evacuation on the grounds that stress, both physical and mental, and the cost of reconstruction of lives in a new location are similar for both groups. Against this is the possibility that someone may move out as a voluntary evacuee with an eye to obtaining a compensatory payout.

Sacrificing a career

Not surprisingly, evacuation of any form typically imposes costs on evacuees' careers. When Mari, in her mid 40s, decided to evacuate from Fukushima City to Nagoya City, she was obliged to give up her job as a school nurse. The family moved when the husband was transferred to a branch office of his company in Nagoya.

> To be honest, it was so hard for me when I had to decide to evacuate since it meant giving up my career. I really wanted to continue my job. At first, I proposed that my husband should move to Nagoya for his work while I and our child remained in Fukushima. Had circumstances been normal I think this would have been acceptable. At that time, circumstances were anything but normal.

Unlike in Akiyo's case, Mari and her husband did not have any relatives in Nagoya, but they did not see that as a deterrent. At least Mari decided to consider being in a new, foreign place as a positive experience.

In another case, Eri and her husband, both freelance engineers in their early 50s, had few constraints on their decision-making. They chose to evacuate, taking their youngest son, who, unlike their other children, still lives with them at home. The husband said that he could get a job in Tokyo as a contract employee at an IT-related company. Eri organizes the "Hoyo to Hawaii" project for children in Fukushima, a recuperation program for child victims of the nuclear disaster. Their family income is lower than before.

Hope within displacement

Despite the significant hardships that many evacuees face in their destinations, I have also encountered interviewees who were seeking to rebuild a sense of confidence and independence. As an example, since Kumi (mentioned above) moved from Fukushima, she has become actively involved in an evacuee network for mothers with children in Tokyo. She told me that she was interviewed by several newspapers, and that the more she spoke about her idea concerning the Fukushima NPP in public, the more a distance grew between herself and her husband. Her husband does not want to move from Otama Village permanently, but she repeatedly insists her priority is to keep her daughter in a safe and secure environment.

Their relationship has worsened and she is now facing a broken family life. Even so, Kumi views the situation as an opportunity to define her own existence.

The same applies to some other informants. Eri, previously mentioned, organizes the recuperation project for children, driven by the sense of mission. Similarly, Kumi organizes a kindergarten group in Sagamihara City with her friends. Yuka, also mentioned above, has become a professional singer-song-writer. She talks of an initially unstable life after leaving Iwaki City, living a kind of hand-to-mouth existence, using up her savings. For Yuka, evacuation from her hometown was equivalent to an escape from a reality, as if she was hiding from something. She did not think that she was able to contribute anything to her hometown. Two years on, she was gradually gaining confidence by producing songs. With support from fellow artists and friends, she made a music CD. Yuka spoke about her experience as an evacuee to her audience. She wishes to stay in Tokyo at least until her two daughters go to senior high school, and she may become a permanent resident of Tokyo.

Some researchers view the situation in the immediate aftermath of the Fukushima nuclear accident as near hopeless. Kuchinskaya, reporting three months after the accident, pointed out that, given Japan's techno-scientific capabilities, the authorities "are able/should be able to provide food delivery services that are routinely tested for radiation" (Kuchinskaya 2014: 162). In reality, the younger generations of Fukushima are not always convinced by this. For example, in June 2011, six young mothers living in Koriyama City set up an association for mothers who wish to share their anxieties about radioactive contamination. The group, called *3As—anshin* (security), *anzen* (safety), and *akushion* (action)—tested locally grown food and locally caught fish for themselves to check levels of radiation, and they later hosted sale events for produce brought in from outside the area. Such activities are likely to produce new social actors, both evacuees and non-evacuees, who are vital to the nation (Beck 2011).

Conclusions

In this chapter I have described selected experiences of Fukushima residents displaced by the nuclear accident that took place on March 11, 2011. Their narratives offer insights into how ordinary people act in catastrophic and highly uncertain situations and how they manage the contingency. I have learned that people voluntarily evacuated following the disaster for three overlapping reasons: first, the wellbeing of their children; second, concern for their own safety; third, the governmental policy regarding free housing provision. The fear of radiation contamination and its effects has been amplified by the government communicating information poorly. In particular, immediately after the accident, a safety level of exposure to radiation was announced in a language that was neither clear nor particularly helpful. My informants all thought that evacuation would be temporary; they did not foresee life away from home for three years and counting.

Beyond these broadly generalized observations, the narratives of the evacuees highlight a wide variety of stories of adversity. Some decisions were made more or less independently, while others were influenced by surrounding voices and emotions. Some evacuees took actions in the immediate aftermath of the accident, while others did so later on. As a result of relocation, some face dysfunctional family situations. All of them have experienced disruptions in their financial and social wellbeing, but the ways in which they cope with these disruptions vary widely. Some seek new lives where they currently are, or somewhere else, while others wish to rebuild their life by returning to Fukushima. My informants, especially mothers with children, maintain social relations and links with their original homes in Fukushima. This implies that they have a place to return to at a later time. However, a fear of radioactive contamination still constrains their lives and decisions. In short, most of them are still in an undecided state.

It is not easy to suggest simple solutions for resolving all of the challenges and problems that these voluntary evacuees face. I believe that policies and support for evacuees must start from the understanding of these concrete realities. Currently there is a vast gap between the timescale in which national and local governments want to claim that reconstruction has been completed and the timescale that evacuees feel they need before they can consider that their lives have been reconstructed. At this point I only hope that wherever they live and whatever their circumstances are, evacuees will have a new consciousness and the courage for a life ahead.

Acknowledgments

This work was supported by JSPS KAKENHI Grant Number 25220403. I would like to express my deep gratitude to Professor Mitsuo Yamakawa for allowing me the opportunity to contribute to this book, and to Professor Daisaku Yamamoto at Colgate University, who gave me many useful suggestions and comments. Finally, I would like to thank Brian Williams for proofreading my writings.

Note

1 Below is the list of interviewed evacuees, cited in the text, in order of appearance. All the names are pseudonyms, followed by their hometowns and the locations and dates of the main interviews.

- Kazuko from Fukushima City, in Tokyo, 17 October, 2014
- Yuka from Iwaki City, in Tokyo, 04 December, 2014
- Akiyo from and in Fukushima City, 15 October, 2014
- Kumi from Otama Village, in Kanagawa, 21 October, 2014
- Junko from Motomiya City, in Tokyo, 24 January, 2015
- Mr Yoshi from and in Fukushima City, 07 November, 2014
- Akemi from Koriyama City, in Tokyo, 18 December, 2014
- Tomo from Kawamata Town, in Tokyo, 14 May, 2015
- Mari from Fukushima City, in Tokyo, 27 February, 2015
- Eri from Koriyama City, in Tokyo, 12 December, 2014

References

Beck, Ulrich. 2011. "Kono Kikai ni" [On This Occasion]. In *Risukuka suru Nippon Shakai: Ulrich Beck tono taiwa* [Risk to Japanese Society: A Conversation with Ulrich Beck], edited by Ulrich Beck, Munenori Suzuki, and Midori Ito, 1–12. Tokyo: Hosei University Press. [In Japanese.]

Fukushima Prefecture. 2016. "Current Conditions and Changes in Evacuation to Other Prefectures." Accessed February 16, 2016. www.pref.fukushima.lg.jp/uploaded/attachment/149711.pdf. [In Japanese.]

Ikeda, Yoko. 2013. "The Construction of Risk and the Resilience of Fukushima in the Aftermath of Nuclear Power Plant Accident." In *Japan Copes with Calamity: Ethnographies of the Earthquake, Tsunami and Nuclear Disaster*, edited by Tom Gill, Brigitte Segar, and David H. Slater, 151–75. Bern: Peter Lang.

Imai, Akira. 2014. *Jichitai Saiken: Genpatsu Hinan to Idou Suru Mura* [The Reconstruction of Local Government: Moving Villages and Evacuation Following the Nuclear Accident], Tokyo: Chikuma Shobo. [In Japanese.]

Kashimura, Aiko. 2011. "2010 Nendai no Nippon ni okeru Kojinka to Bekku no Riron" [Individualization and the Theory of Beck in 2010s Japan]. In *Risukuka suru Nippon Shakai: Ulrich Beck tono taiwa* [Risk to Japanese Society: A Conversation with Ulrich Beck], edited by Ulrich Beck, Munenori Suzuki, and Midori Ito, 53–69. Tokyo: Housei University Press. [In Japanese.]

Kuchinskaya, Olga. 2014. *The Politics of Invisibility: Public Knowledge about Radiation Health Effects after Chernobyl*, Cambridge, MA: MIT Press.

Seki, Reiko, and Yuka Hiromoto, eds. 2014. *Tosu no Tsumugi-Mou Hitotsu no Yutopia* [The Narratives of Evacuees in Tosu City—For a Further Utopia], Tokyo: Hinsensha. [In Japanese.]

Yamakawa, Mitsuo. 2013. *Gensaichi Fukko no Keizai Chirigaku* [The Economic Geography of Restoration after Disaster], Tokyo: Sakurai Shoten. [In Japanese.]

Yamane, Sumika. 2013. "Genbatsu Jiko ni Yoru Boshihinan Mondai to Sono Shien" [Problems and Support of Mother and Child Evacuees Resulting from the Fukushima Nuclear Accident]. Annual Report at Yamagata University, 10: 37–51. Accessed February 18, 2016. www-h.yamagata-u.ac.jp/wp-content/uploads/2014/09/nenpou10_03.pdf. [In Japanese.]

Yamashita, Yusuke, Takashi Ichimura, and Akihiko Sato. 2013. *Ningen Naki Fukko: Genpatsu Hinan to Kokumin no "Furikai" o Megutte* [Revitalization without People: Evacuation following the Fukushima Nuclear Accident and the Lack of Understanding from the Nation], Tokyo: Akashi Shoten. [In Japanese.]

Yoshimura, Ryoichi. 2015. "Jishuteki Hinansya Kuiki Gai Hinansya to Senzaisya no Songai" [Damage Compensation to Voluntary Evacuees and Residents]. In *Fukushima Genpatsu Jiko Baisho no Kenkyu* [A Study of Compensation following the Fukushima Nuclear Accident], edited by Awaji Takehisa, Ryoichi Yoshimura, and Masafumi Yokemoto, 123–39. Tokyo: Nippon Hyoron Sha. [In Japanese.]

6 How safe is safe enough?

The politics of decontamination in Fukushima

David W. Edgington

I hate the word reconstruction. The government is using this to hide the true situation of people's lives in Fukushima.

(Hiroko Aihara, journalist, Fukushima City, June 2014)

The government thinks that by concentrating on the recovery of infrastructure and housing and the decontamination then they have done their job. But they have not done their job of looking after the safety of residents very well.

(Akemi Shima, concerned mother, Fukushima City, June 2014)

Introduction

Since March 2011, the residents of Fukushima have been living with increased levels of radiation released from the Fukushima Daiichi Nuclear Power Plant (NPP). This chapter discusses the politics of decontamination programs carried out in the three years up to 2014 through a case study of Fukushima City, a mixed urban and rural community, one that was not affected by formal evacuation orders but where residents have had to endure low levels of radiation. The study documents both formal government programs as well as the inimical reactions of non-governmental organizations (NGOs), not-for-profit organizations (NPOs), and local citizens.

The policies and practices that drive decontamination are rooted in decisions made regarding the health and safety of affected residents, particularly evacuees. But even at relatively low levels of contamination, many commentators have argued that there is no absolute "safe level" with regard to radiation exposure. Therefore, government policy has had to determine *acceptable* levels of safety. In other words, it has to address "how safe is safe enough?" (Fischhoff *et al*. 2000).

As well as the opinions of experts on matters of radiation, mainly those of scientists and technocrats, citizen viewpoints—concerning their own "perceived risk"—are also important, especially regarding the health of young children. Consequently, in this study I have tried to capture the concerns of a number of local residents in the first three years after the NPP accident and their response to a number of controversial issues that emerged in decontaminating Fukushima City, including claims that the targeted reduction of radiation levels were not set sufficiently low to ensure acceptable safety, inadequate oversight of the decontamination program, problems concerning storage of tons of contaminated debris, and the lack of responsiveness to requests by citizens for more thorough decontamination of radiation "hot spots" in specific areas.

I have been struck by the "cavernous disconnect" that has emerged between the Tokyo technocrats on the one hand—who argue that standards regarding evacuation, decontamination, and return should be calculated in terms of a quantifiable risk–benefits analysis—and the residents and NPOs of Fukushima on the other, several of whom argue that ethics and human rights demand the full evacuation of contaminated residents, or at least the women and children.[1] This study focuses therefore on questions surrounding how formal government approaches to radiation risk to health were formulated after March 2011, how the decontamination programs of local governments have been carried out, and what factors have shaped the risk perceptions of citizens and NGOs.

The research is based on fieldwork carried out in Fukushima during 2013 and 2014, analysis of various Japanese and overseas reports dealing with the nuclear disaster, radiation emissions, and government decontamination programs, and discussions with local residents, academics, and policy makers in Fukushima Prefecture.[2] The chapter first discusses theoretical concepts of risk management and the perception of risk and indicates the key threshold levels of radiation exposure used by the Japanese government and local governments in devising programs to mitigate the health impacts of radiation in Fukushima. There then follows a brief chronology of events since March 2011, indicating the spatial variation inherent in government decontamination programs, together with a discussion of various decontamination issues in Fukushima City (part of the national government's designated Intensive Decontamination Survey Area), such as calls for evacuation in the district of Watari, protection of young children at schools, the monitoring of radiation "hot spots" around the city, and methods to dispose of radioactive waste. The chapter concludes by assessing the divide between Fukushima residents and technocratic-generated policies in terms of risk communication and trust and the degree to which this has been (or even could be) mediated by the actions of local government.

Risk analysis and low-level radiation

> Life is full of risk. Risk is not the antonym of "safety," nor does it mean simply "danger."
>
> (Working Group on Risk Management of Low-dose Radiation Exposure, Office of the Deputy Chief Cabinet Secretary, December 2011)

As implied by the above quote, official debates about public anxiety resulting from the Fukushima nuclear accident have revolved around how to evaluate the risks of low-dose-radiation contamination of soil, residential areas, and farmland, and how to communicate these risks to the general public (Hirakawa and

Shirabe 2015). In this regard, Palenchar and Heath (2006) note that there are at least three theoretical options to guide the way in which risks are calculated, controlled, and communicated. The first may be called the "scientific positivism" approach, whereby data and methodologies of scientists dominate efforts to ascertain the degree of risk. Once a decision has been made, usually in terms of calculating the "risk–benefit" of any hazardous activity, then a standard is set (for example, a standard regulating the appropriate level of radiation exposure in a particular community) and elite technocrats manage and communicate the risk on behalf of the public (Wilson and Crouch 2001). A second approach, favored by psychologists such as Slovic (2000a), involves constructivism or relativism, which assumes that everyone's opinions have equal value, so that no individual's opinion about risk, or perceived risk, is better or worse than anyone else's. A third option is dialogue, whereby risk assessment and decisions about managing risk are made through collaborative decision-making processes. Here, scientific opinions are integrated into opinions about risk that are vetted by key public values (Renn 2009).

In framing debates about radiation exposure, some scientists, regulators, and members of risk-producing industries (for example, the nuclear -power industry) scoff at any risk-assessment and decision-making approach other than the first option. In the Fukushima context, Hirakawa and Shirabe (2015) draw attention to this attitude by describing the "scientific deficit model" of risk communication in Japan, used widely by the government after March 2011, whereby, in communicating to the public, bureaucrats and scientists assumed that a lack of clear and accurate knowledge about low-level radiation was the major source of people's anxiety and the possible long-term health impacts, rather than a wider mix of both material and perceived problems in the wake of the NPP accident. Moreover, these authors argue that official explanations of radiation issues since the nuclear accident were designed not just to dispel people's unease but to marginalize any public concern by presenting the "correct" scientific knowledge about the underlying risks involved. However, contrasting the "mainstream" views of "expert" scientists and other government technocrats, other risk-communication practitioners contend that the public's acceptance of various risks after the Fukushima accident requires the collective judgment of various stakeholders in the community and is more successfully crafted through dialogue (Fukushima Action Research on Effective Decontamination Operation 2013). Indeed, many studies have found that people feel that they enjoy more control when they have a say in the formulation of the factors that seem to lead to the control of risks (Renn 2009).

Favoring an approach to risk that focuses on the latter approach—one that takes into account public perceptions and concerns—Fischhoff and his colleagues have initiated "expressed-preference research," which involves measuring a wider array of attitudes rather than merely weighing the measurable benefits in the effort to ascertain tolerable risk levels (Fischhoff and Kadvany 2011). They found that laypeople's risk ratings, unlike those of experts, are not influenced merely by quantitative fatality estimates, but also by their judgments of several qualitative factors. In other words, "risk is not just about risk," meaning that a wider set of social, cultural, and psychological factors should also be incorporated into risk-analysis decisions. Of particular note, studies have shown that the public evaluates a particular risky activity or technology as more "dangerous" if it is involuntary, unfamiliar, unknown, uncontrollable, controlled by others, unfair, memorable, dreaded, acute, focused in time and space, fatal, delayed, artificial, and undetectable (Palenchar and Heath 2006).

Of course, this describes well the many controversies surrounding perception of risks from radiation and nuclear power, together with the politics of disposing of nuclear waste (Slovic 2000b; Slovic *et al.* 2000). Also of concern to Slovic (2000c) is that, while a certain amount of paternalism seems unavoidable when formal regulatory bodies undertake risk management, any lack of trust shown by the public towards scientists and government regulators is a critical factor underlying controversy over technological hazards. Indeed, the playing field is tilted towards "distrust," especially when "negative" or trust-destroying events occur, such as unexpected power-plant accidents in otherwise safe operations of the nuclear industry. Moreover, in his studies of risk perception, Slovic (2000c) found that women and non-white men (in a North American context) tend to perceive greater risk, because they have less control over certain activities and technologies and benefit less from them. Indeed, risk perception seems to be related to the individual's power to influence decisions about the use of hazards. His research therefore paints a portrait of risk perception influenced by a complex interplay of psychological, social, and political factors.

As indicated by Hirakawa and Shirabe (2015), perhaps the most critical policy debate in Fukushima post-March 2011 concerned the risk of low-dose radioactive contamination of food, water, soil, and tsunami debris. In terms of radiation, past studies have shown a rise in cancer rates in populations exposed to a dose of 100 mSv/year or radiation;[3] yet they reveal much less about the situation in Fukushima, where lower doses will continue for many years.

How dangerous to humans are low doses of exposure to radioactivity? There is no clear definition of "low-dose exposure," but some commentators refer to a level of 200 mSv/year or less (Streffer *et al.* 2004). Table 6.1 indicates the dose implications of various levels of radiation. Comparisons are often made between different risks from estimated radiation exposure. For instance, the radiation dose from each X-ray or medical procedure, such as CT scans, is between 50 μSv to 3–9 mSv. Moreover, one round-trip air flight between Tokyo and New York exposes passengers to roughly 0.2 mSv of radiation, due to increased cosmic rays from flying at high altitudes, and this may rise to a yearly dose of around 9 mSv for pilots and crew flying between Japan and North America (see Table 0.1 in the Introduction). However, these simple comparisons between risks from radiation exposure are not necessarily appropriate in framing discussions of post-NPP-accident radiation policies, for, as we have seen, people evaluate risks taken voluntarily in a very different way from those imposed by technology such as nuclear power (see also Hirakawa and Shirabe 2015).

In relation to these various levels of radiation, what standards did the Japanese government set in choosing levels of risk concerning post-accident decontamination and for what purpose? As suggested earlier, a key indicator of managing hazards and gauging community risk after NPP accidents is whether or not a person is exposed to more or less than 100 mSv/year, which is the level below which medical science cannot prove a higher risk for cancer (see Table 0.1 in the Introduction), based on research conduced in Hiroshima, Nagasaki, and Chernobyl (Allison 2009). While this standard might be said to be the "conventional viewpoint" of radiation in medical science, certain commentators take a position that there is no threshold of a "non-hazardous level."[4] Even levels of radiation lower than 100 mSv/year may increase the risk of long-term threats to health and the possibility of contracting cancer and other diseases. Moreover, children up to the age of 15 years are said to be two to three times more sensitive to the effects of radiation than adults (Meister 2005).[5]

As we shall see in the following section, in setting standards that guided decontamination policies the Japanese government consulted with the International Atomic Energy Agency (IAEA) and also relied on recommendations from the International Commission on Radiological Protection (ICRP), which provided a reference "band" of radiation doses from 1–20 mSv/year for the Japanese government to consider for resettling people in contaminated areas. The ICRP classifies radiation-exposure situations into three different types: "emergency," "existing," and "planned" (Wrixon 2008).

Figure 6.1 indicates that, following a nuclear accident (the "emergency exposure phase"), the ICRP recommended a radiation dose range of between 20–100 mSv in setting targets for the design and carrying out of measures for protection from radiation. As indices for efforts to gradually improve the situation in following an emergency, the ICRP recommended that interim reference levels should be established, with an initial threshold of 20 mSv/year, and efforts made toward improvement based on a long-term annual target of 1 mSv. For long-term "planned" exposure situations, establishment of "dose constraints," rather than use of reference levels within the range of an annual 1 mSv or less, has been recommended by the ICRP, depending on the situation regarding the exposure status of the general public (Wrixon 2008). These principles and how they have guided government programs (described shortly) are summarized in Figure 6.1. Essentially, the Japanese government chose a 20 mSv/year standard as a threshold to gauge whether or not to evacuate residents from hazardous zones and whether to allow them to return. In terms of a long-term goal, the government chose the much lower 1 mSv/year level, which may be called the "clean for all practical purposes" level (Safecast 2013).

Having outlined the quantitative aspects of low-radiation-exposure risk and the various standards chosen by the national government, the remainder of the chapter discusses spatial variations in decontamination policies adopted for Fukushima Prefecture and eight neighboring prefectures, together with reactions to decontamination programs implemented in Fukushima City, in terms of the risk perceived by NGOs and residents.

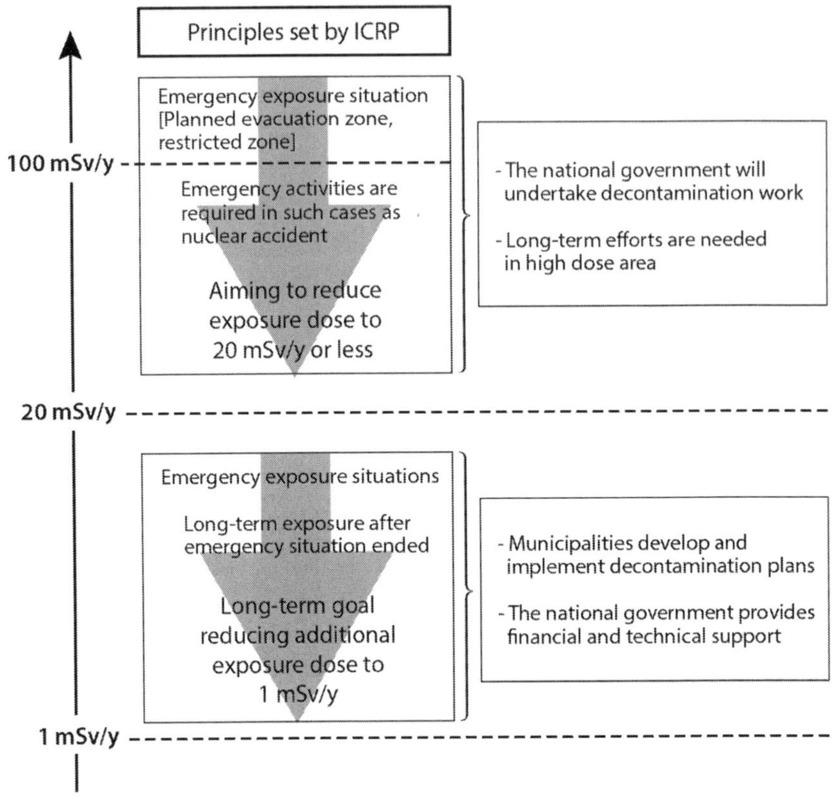

Figure 6.1 Basic approach to decontamination work.

Source: Author, based on Ministry of Environment (2013b).

The geography of decontamination

In the immediate chaos of evacuating an estimated 84,000 people from the stricken Fukushima Daiichi NPP in March 2011, there was very little information to begin with on just how much radiation had leaked into the atmosphere and which areas of the prefecture and beyond had been contaminated (Suzuki and Kaneko 2013). Official aerial mapping of contaminated areas by the Japanese government was not publicly released until May 2011 (Sanada *et al.* 2014). However, as more Geiger counters and other devices to detect radiation found their way into Fukushima Prefecture, citizens' groups, municipal administrations, and individual residents began to measure local radiation levels themselves. The worst radiation readings were taken northwest of the Fukushima NPP, on the sides of mountains and hills facing the plant, along an approximately 50 km plume, apparently because winds blew radiation in that direction after the most intense concentrations of radiation were released during the hydrogen explosions from the NPP on March 11–12 and were then brought to the ground by the mountains and weather systems around the mountains in the days following (Figure 2.2 in Chapter 2).

Faced with the knowledge that areas of Fukushima were contaminated, many communities outside the formally evacuated areas took decontamination into their own hands. Volunteers joined local residents and farmers and began removing the top few centimeters of radioactive soil from schoolyards and parks where young children played, as well as rice fields (Brumfiel and Fuyono 2012). With the help of independent researchers, many local governments removed contaminated soil and conducted other clean-up operations themselves, using municipal workers and local residential volunteers (Onishi and Fackler 2011). In the summer of 2011, the cities of Fukushima, Date, and Minami-Soma announced ambitious plans to decontaminate their own areas, starting with schools and other parts of towns frequented by children (Mahr 2011). As the year wore on, Fukushima Prefecture residents and evacuees faced extreme anxiety about radiation health risks as well as numerous psychological and social burdens resulting from the limits of lifestyle associated with protection measures, such as having to keep young children indoors (Hanai and Lies 2014). Although residents and local governments started voluntary decontamination, it became apparent that this was ineffective in the absence of expert planning and advice.

Accordingly, the national government devised its own formal strategy for decontamination and began analyzing which areas of Fukushima Prefecture and the wider Tohoku region should be given priority for decontamination. In December 2011, the government passed the Act on Special Measures Concerning Radioactive Contamination. Following the enforcement of this legislation in January 2012, a framework and guidelines for decontamination operations was released which covered approved methods for surveying and measuring the degree of contamination, as well as measures for decontamination and guidelines for the collection, transport, and storage of removed soil and other radioactive waste. In line with the ICRP recommendations, the government determined that decontamination of areas with over 20 mSv/year annual cumulative dose of airborne radiation would be the responsibility of the national government, and areas with readings below this level would be carried out by municipal administrations. For all areas where the annual radiation dosage exceeded 1 mSv/year, excluding naturally occurring radiation, the government set a long-term goal of reducing the level through decontamination to below 1 mSv/year (Ministry of Environment 2013a, 2013b).

To operationalize these goals, the national government determined two types of decontamination zone. The first, the "Special Decontamination Areas," comprised the 20 km restricted evacuation zone around the Fukushima Daiichi NPP, together with municipalities evacuated later in 2011 and located up to 60 km northwest of the stricken plant (Figure 6.2). In this zone, decontamination was implemented by the national government and covered the municipalities of Naraha, Tomioka, Okuma, Futaba, Namie, Katsurao, and Iitate, together with some areas of Tamura, Minamisoma, Kawamata, and Kawauchi (Figure 6.2). The second type of zone, called "Intensive Contamination Survey Areas," covered not only "unevacuated"

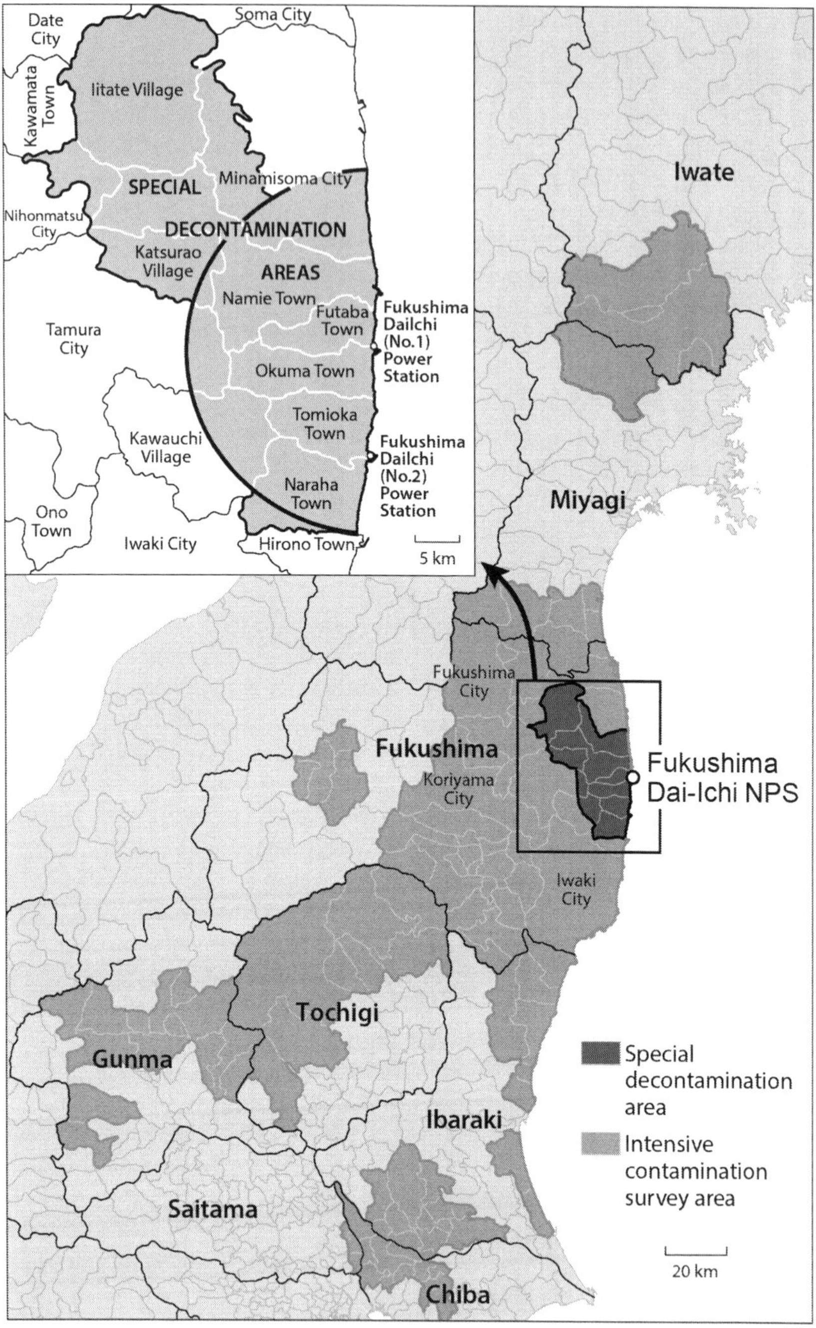

Figure 6.2 Special Decontamination Areas.

Source: Author, based on Ministry of Environment (2013b).

towns in Fukushima but over 100 municipalities in eight prefectures where additional radiation doses over 1 mSv/year were determined (Figure 6.2). In these areas, decontamination was left to individual municipalities, supported by national funding and with prefectural governments providing technical advice (Ministry of Environment 2013a, 2013b).

Basically, the restricted areas of Fukushima that were affected by evacuation orders in early 2011 were taken over by the national government in lieu of any effective local administration, due to the emptying of population and municipal offices. These areas had the highest ground contamination after the nuclear accident, and the locations that were the most difficult to decontaminate lay within this zone. The Japanese government acknowledged this problem and started conducting trial decontamination programs within the Special Decontamination Areas in November 2011 under the control of the Japan Atomic Energy Agency (JAEA), with an initial estimated budget of 10.9 billion yen (about 100 million US dollars). The JAEA assigned the project to joint ventures led by three major construction companies—Taisei-Kensetsu, Obayashi-Gumi, and Kajima-Kensetsu—which tested various technologies to clean up radioactive materials in the 11 cities, towns, and villages whose citizens remained evacuated.[5] These companies came up with novel ideas to remove decontaminated radioactive cesium based on existing technologies. For example, scouring cesium from roads with a high-pressure water jet was thought initially to be insufficient because contaminated water would simply spread across the pavement. However, engineers modified this system to recover the decontaminated water and then purified and recycled it (Brumfiel and Fuyuno 2012). The results of these experiments guided the large-scale-decontamination program that the government began the following year (Ministry of Environment 2013a, 2013b).

Prior to focusing on a case study of decontamination in Fukushima City, it is worth pointing out certain features and implications of the government's standards and procedures for its program in the Intensive Contamination Survey Areas. First, while health specialists in radiation continue to debate the risks of doses present in Fukushima, it seems even the most outspoken anti-nuclear activists in Japan have been willing to accept an extra 1 mSv/year and acknowledge that adding 1 mSv/year to background radiation levels would keep decontaminated areas within the typical range of normal background radiation worldwide, which is about 1–3 mSv/year (not including the so-called high natural background radiation areas such as Ramsar, Iran and Kerala, India, which are many times higher) (Safecast 2013).[6]

A second dimension to the government's choice of radiation levels for different parts of Fukushima is the operational level of 0.23 μSv/hour, measured by dosimeters "on the ground," as a way of indicating whether the annual standard of 1 mSv/year had been achieved by decontamination. This was developed by the Japanese Ministry of Environment (MOE) in a mathematical model that converted the ambient dose rate in μSv/hour to an annual dose. The formula below describes how the model was produced (Embassy Science Fellows Team 2013).

> {[0.23 (measured air dose rate) – 0.04 (natural background radiation dose rate) μSv/h] x [8h+16h x 0.4 (shielding factor due to staying indoors in wooden housing)]} x 365 days/1000 = 1 mSv/year.

Essentially, the model is an estimation of the annual dose that would be received by a "reference" individual living in an area with a given measured airborne dose rate. This estimation assumes 8 hours of outdoor and 16 hours of indoor activities. The government dose-rate calculations therefore are not based on simple arithmetic but assume dose-reduction benefits from shielding provided by buildings and reflect an attempt to account for time spent indoors. Moreover, to determine the amount of radiation coming from contamination, the pre-existing background radiation needs to be subtracted from whatever background reading is obtained in any one spot, here assumed to be the 0.04 μSv/hour for the Fukushima average (Safecast 2013).

The 0.23 μSv/hour standard has been important for decontamination programs in the Intensive Decontamination Survey Areas as the national government will only subsidize municipalities when radiation exceeds it. Brasnor (2012) reports that some residential families and farmers in Fukushima have cleaned their properties themselves in order to bring their levels even lower, going as far as destroying gardens and trees, even though the effectiveness is not clear. At any rate, municipalities made huge efforts to undertake the unprecedentedly large project of clearing up radioactive materials scattered across a huge area. Despite various problems, such as insufficient human resources, knowledge, and experience, some progress has been made since March 2011. While decontamination of public facilities (for example, schools and parks) advanced well, progress in decontaminating houses up to the end of 2014 was more varied due to diverse conditions and approaches. Some cities that undertook decontamination shortly after the accident accumulated valuable knowledge about relevant technologies, modes of communication with residents, and consensus-building (interview with Mr Keiihi Miho, former mayor of Nihonmastu City, Fukushima, June 2014). However, this knowledge was not often shared with other municipalities as most technical knowledge was accumulated within the various construction firms that carried out the decontamination (Safecast 2013).

City of Fukushima case study

Fukushima City has the third largest population in Fukushima Prefecture, with around 290,000 persons. There are 19 districts within the city, many of them in rural hilly locations outside the central urban areas (Figure 6.3). Figure 6.4 indicates the nature of the terrain around the city, together with the nearly 800 monitoring stations that the city established in 2011 to measure airborne radiation. (These stations are different from the fixed monitoring posts (395 posts in Fukushima City) installed and operated by the Nuclear Regulation Authority). Even though the city is located about 60 km west of the Fukushima Daiichi NPP, the eastern part of the city recorded levels of cesium-134 and 137 contamination over 2.0 μSv/hour (see Figure 6.5). Table 6.1 indicates that May 2011 radiation air dose levels ranged from 0.26 μSv/hour in the Tsuchiyu Onsenmachi district in the west of the city (about 1.2 mSv/year) to 2.23 μSv/hour in the Watari district and 2.24 μSv/hour in the Onami district (about 11.6 mSv/year) located in the eastern part of the city (Table 6.1). Table 6.1 also indicates that a wide range of radiation levels was recorded *within* each district in 2011.

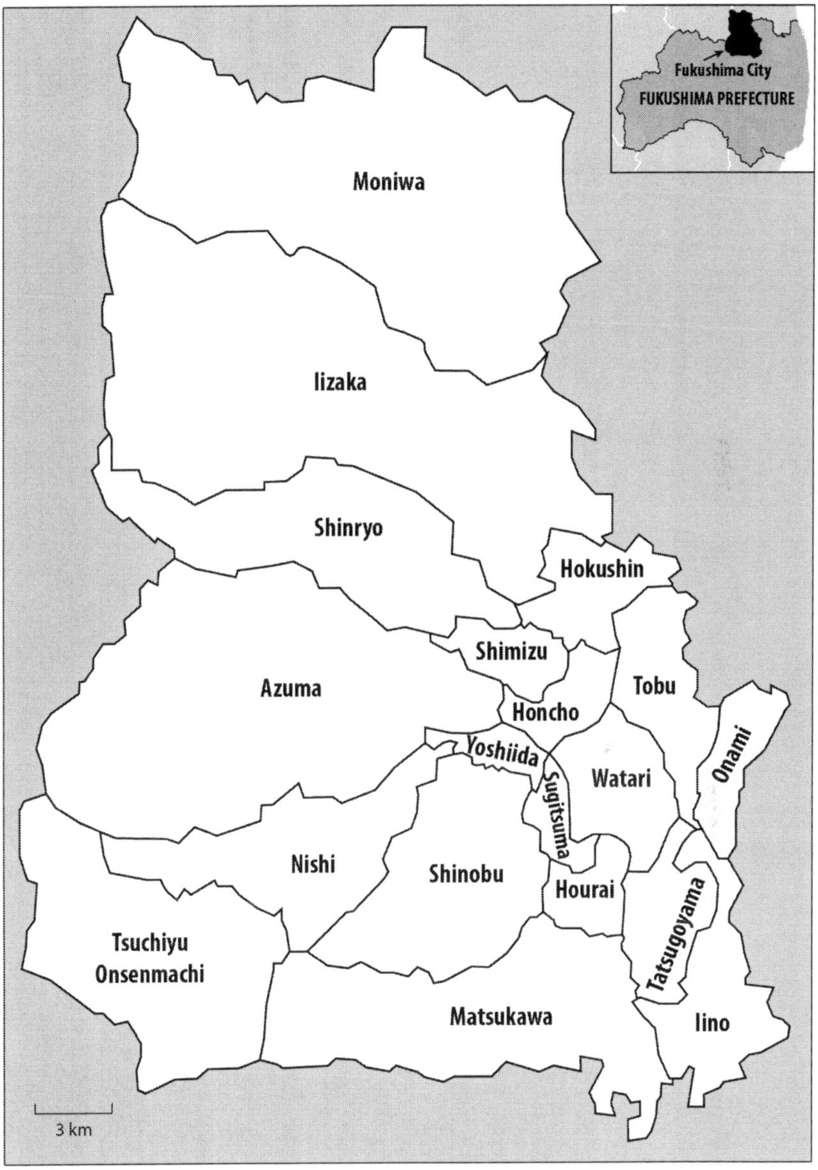

Figure 6.3 The 19 districts of the City of Fukushima.

Source: Author, based on data provided by the City of Fukushima.

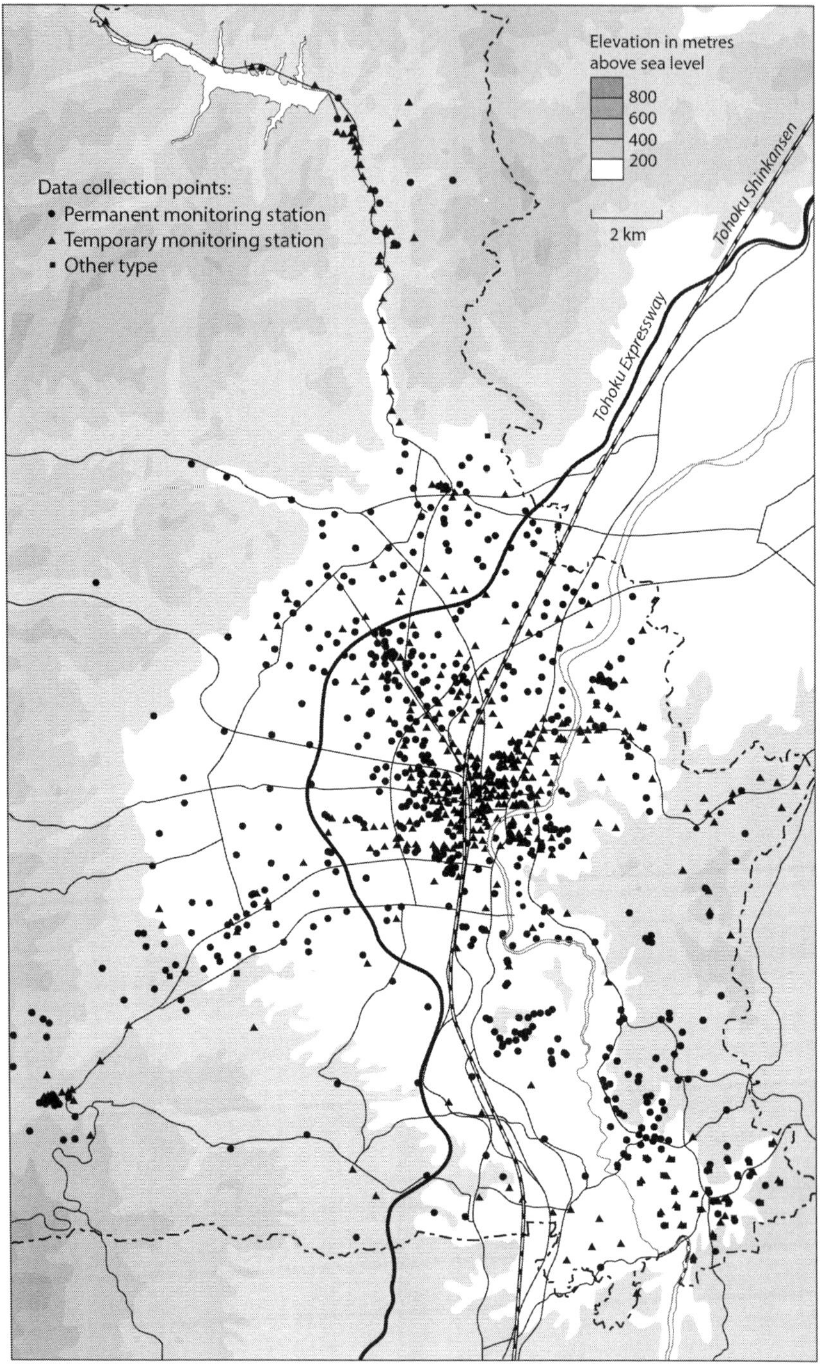

Figure 6.4 Distribution of radiation-measuring monitors in the City of Fukushima, 2011.

Source: Author, based on data provided by the City of Fukushima.

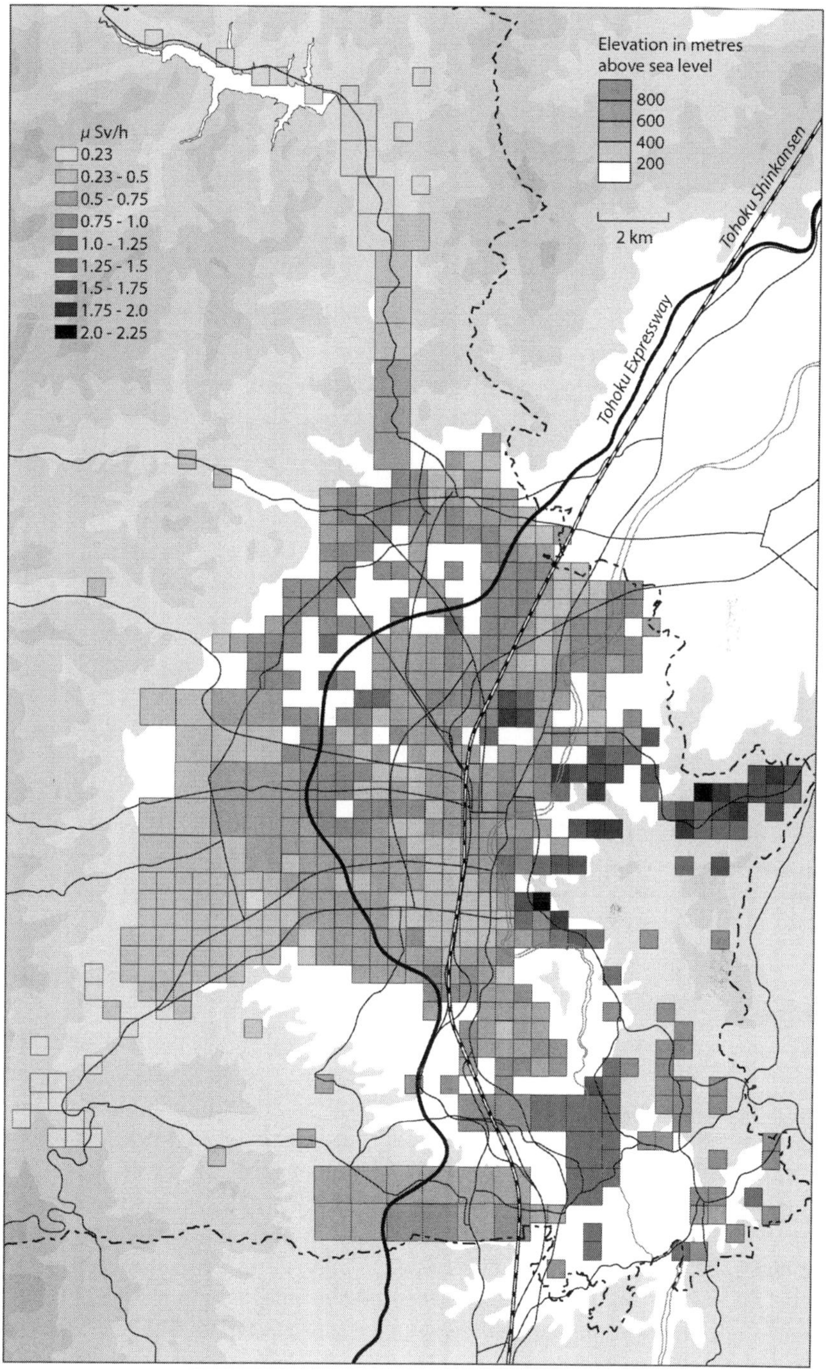

Figure 6.5 Distribution of radiation in the City of Fukushima, March 2012.

Source: Author, based on data provided by the City of Fukushima.

Note: 0.5×0.5km and 1.0×1.0km grids.

Table 6.1 Changes of amount of ambient dose by city district (unit: μSv/h)

District	*Air-borne radiation May 2011*	*Average May 2011*	*Average March 2013*	*Average March 2014*	*Reduction 2011–14 (%)*	*Reduction 2013–14 (%)*
Chuou	0.71–3.32	1.59	0.51	0.32	79.9%	37.3%
Watari	1.02–4.05	2.23	0.86	0.52	76.7%	39.5%
Sugitsuma	0.42–2.02	1.17	0.34	0.22	81.2%	35.3%
Hourai	1.03–2.22	1.55	0.52	0.30	80.6%	42.3%
Shimizu	0.71–2.95	1.80	0.51	0.36	80.0%	29.4%
Tobu	0.55–3.00	1.60	0.77	0.48	70.0%	37.7%
Onami	1.25–3.87	2.24	0.97	0.65	71.0%	33.0%
Hokushin	0.77–2.73	1.43	0.53	0.36	48.8%	32.1%
Yoshiida	0.58–2.20	1.19	0.40	0.28	76.5%	30.0%
Nishi	0.26–1.14	0.63	0.29	0.21	66.7%	27.6%
Tsuchiyu Onsenmachi	0.11–0.40	0.26	0.14	0.08	69.2%	42.9%
Shinryo	0.74–2.64	1.63	0.59	0.43	73.6%	27.1%
Tatsugoyama	1.19–2.33	1.76	0.81	0.51	71.0%	37.0%
Iizaka	0.42–2.13	1.05	0.56	0.40	61.9%	28.6%
Moniwa	0.12–1.05	0.33	0.16	0.11	66.7%	31.3%
Matsukawa	0.52–2.08	1.16	0.69	0.42	63.8%	39.1%
Shinobu	0.56–1.75	0.91	0.40	0.28	69.2%	30.0%
Asuma	0.32–1.78	1.15	0.41	0.31	73.0%	24.4%
Iino	0.76–6.65	1.58	0.75	0.49	69.0%	34.7%
Total city	—	1.33	0.56	0.37	72.2%	33.9%

Source: Author, based on data provided by the City of Fukushima.

As noted earlier, Fukushima City was not designated as an evacuation zone. However, even though factories, shops, and farms have been active since the accident, it is estimated that around 6,000 people voluntarily left the city and were still living away from their homes two years after the NPP accident (*Asahi Shimbun* 2013). Due to the urgency of protecting residents, and fearful of a more serious flight of population, the municipal administration began an extensive survey of radiation levels all over the city, followed in the summer of 2011 by decontamination measures by some 3,700 city employees, together with resident volunteers (interview with Mr Yudai Ouchi, Assemblyman, City of Fukushima, June 2014). Due to the importance of farming in the community, the city has also been working with local agricultural cooperatives to understand the implications of contamination for local farmers (interview with Mr Yuta Hirai, Fukushima Consumer Coop Program, May 2013).

In September 2011, the city drew up its own basic principles covering decontamination (the Furusato Decontamination Plan) and then produced a "second edition" in May 2012 in line with the statutory plan and guidelines laid out by the national government (City of Fukushima 2012). As with other municipalities in the Special Decontamination Areas, the program is implemented by the city administration, but costs and standards to be attained in the process, together with long-term goals, are set by the national government. However, the city was able to set its own priorities for implementing decontamination efforts and, perhaps not surprisingly, urgency was assigned to the two districts in the east of the city showing the highest air doses, Onami and Watari. Priority was also given to areas that would be frequently used by children, such as roads, schools, and parks.

Figure 6.6 indicates the broad-scale work plan established in fiscal year 2012, indicating the city's priorities and their resolve to address the decontamination of houses in Onami and Watari immediately, together with public facilities such as schools and parks. However, due to the limitations of the personnel that could be hired for this work, housing decontamination advanced slowly and only 25,000 houses out of around 110,000 in total were planned for decontamination in 2012. To decontaminate a house correctly took from a few days to a few weeks. Topsoil had to be removed from the garden, moss had to be scraped away, and concrete paving had to be power washed (interview with Mr Yudai Ouchi, Assemblyman, City of Fukushima, June 2014).

Based on fieldwork conduced in 2013 and 2014, I observed that the decontamination taking place around Fukushima City was surprisingly low-technology, involving shovels, long-handled brushes, and high-pressure water guns (Table 6.2). In an attempt to reach the long-term radiation goal of 1 mSv/year, thousands of temporary workers have been put to work scrubbing houses and roads, digging up topsoil, and stripping leaves into which invisible cesium particles have

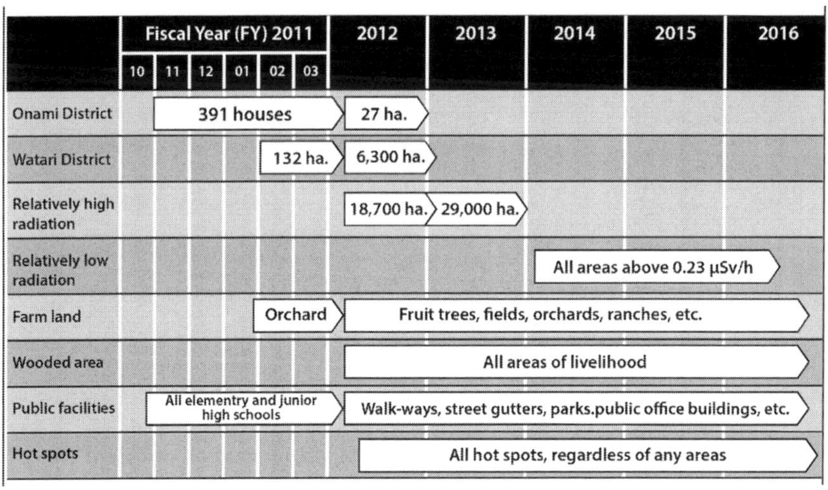

Figure 6.6 City of Fukushima decontamination work plan, fiscal year 2012.

Source: Data supplied by the City of Fukushima.

wormed. In Fukushima City, radiation from the contaminated plume settled on exposed surfaces; these had to be scrubbed and cleaned and topsoil removed. In the outer suburbs of the city, forests were only decontaminated up to 20 m from houses. After falling with dust on to the ground and other surfaces, cesium is no longer in the atmosphere, but over time it has bound with clay contained in soil. As parks and school grounds were stripped of topsoil, the city buried irradiated soil on the sites from where it has been removed. Contaminated dirt was covered in at least 50 cm of clean topsoil, according to city guidelines, a process that officials said had lowered radiation levels by 80 percent in schools (interview with Mr Yudai Ouchi, Assemblyman, City of Fukushima, June 2014).

In many respects, the decontamination program in the city is relatively transparent. The work plan is shown on the city's website, and there are links to a detailed map of radiation levels, an explanation of how decisions to conduct the work are being reached, and timelines for decontamination in each housing district, with achievement goals. There are also clear efforts to inform citizens through neighborhood associations, PTAs at schools, volunteer groups, and companies. A Citizens Decontamination Advisory Corner was set up on the ground floor of the city administration building in 2012.

Over the three years up to mid 2014, the city could point to an overall lowering of radiation levels of around 70 percent on average within its jurisdiction (Table 6.1), although whether this was due to active decontamination programs or (more likely) the natural decay of cesium-134 is unclear.[7] For instance, Table 6.1 indicates that the priority districts of Watari and Onami, where decontamination procedures had been completed in 2014, had only achieved an average decline in

Table 6.2 Decontamination techniques used to remove radioactive material from different objects

Decontaminated item	*Decontaminated technique used*
Eaves and roof gutters	Wiping and high-pressure washing after removing deposited material
Storm-water catch basins	High-pressure washing after removing deposited material
Street gutters	High-pressure washing after removing deposited material
Roofs	Wiping, washing, high-pressure washing
Outer walls	Wiping, washing, high-pressure washing
Gardens and other grounds	Mowing grass, collection of clippings, pruning, surface soil removal, replacing turf, plowing
Parking lots and other paved surfaces	Washing, high-pressure washing, surface-dirt removal (shot blasting), grit-blasting, and so on
School athletic grounds and so on	Surface-dirt removal
Roads (asphalt-paved surfaces)	Washing, high-pressure washing, shaving off

Source: Author.

overall decontamination since 2011 of around 70–5 percent. However, the adjoining districts of Sugitsuma, Hourai and Shimizu, which had average levels of contamination in 2011, recorded higher decreases, of over 80 percent. On the face of it, these figures appear to invalidate the local government's claims concerning the overall effectiveness of the decontamination program. As shown in Table 6.1, average air dose radiation in the city during March 2014 was 0.37 μSv/hour, still above the long-term goal of 0.23 μSv/hour.

On the other hand, whether or not the formal program has proven useful or not, the city attracted the ire of certain NGOs and residents, who were galvanized around a number of decontamination issues.

Concerns of citizens over radiation monitoring

The NGO Human Rights Now claimed that the city administration did not correctly measure radiation, especially where residents themselves had found high levels of radiation. This issue concerned the technical method of measuring ambient radiation levels in particular areas. Using standard procedures, the in-situ monitoring stations set up by the municipality in 2011 were designed to measure at the height of one meter. For some, however, this was unsatisfactory; in particular, residents with children suspected the city government of trying to misrepresent the severity of radiation measurements.

> Residents reported that levels of radiation tended to be higher closer to the ground level, and although parents were concerned with the effect of soil contamination and of radiation close to the ground on children, the government maintained a standard of measuring radiation from one meter from the ground level.
>
> (Human Rights Now 2012: 3)

The city disputed the NGO's claim that it had underestimated radiation contamination and its risks and pointed out that radiation in Fukushima City had steadily diminished over time (interview with Mr Yudai Ouchi, Assemblyman, City of Fukushima, June 2014).

The Watari district

As we have already seen, parts of the city recorded higher radiation levels than others and in some cases were even more contaminated than some of the Special Decontamination Areas closer to the NPP. The Watari district in Fukushima City contains about 6,700 households; around 16,000 people live in this residential area, surrounded by trees and mountains along the Abukuma River. It drew public attention in 2011 when its residents, concerned with high concentration, requested that the city and the Japanese government order a special evacuation of the area using the provisions of the Specific Spots Recommended for Evacuation program introduced by the national government but administered by each municipality

(Friends of the Earth 2013). However, the district was not categorized as an evacuation area as the government determined that while the estimated radiation level in this district was certainly high—with some parts showing readings of 5 to 10 μSv/hour in 2011—it was lower than the 20 mSv/year level that was used to trigger an official evacuation of residents.[9] Nonetheless, contrary to the government's decision, residents claimed they had indeed found "hot spot" areas, where radiation levels exceeded 20 mSv/year, despite the government's own measurements. Meanwhile, as noted earlier, decontamination of residential areas in Fukushima City had been slow to start: at the time of the investigation by Human Rights Now in November 2011, organized decontamination was just underway in the high-radiation-level Onami district, but in the Watari district nothing had been done (Human Rights Now 2012).

After the government's refusal to designate the district, parents in the area established a private organization called Save Watari Kids and continued to conduct their own radiation measurements in places frequented by children, such as the routes to and from schools. Some Watari residents subsequently left the district. However, in the absence of any financial compensation for voluntary evacuees, most families were not able to move from the contaminated area for a variety of reasons, such as work or school. Other NGOs, such as Friends of the Earth, organized weekend "rest and recuperation programs" for children at mountain hot springs located in the western part of Fukushima City, about 30 minutes by car from the Watari district, where recorded radiation rates were much lower. The objective was to reduce the radiation exposure of children and allow them to play freely outdoors, as many were kept indoors by parents fearful of the effects of radiation. One resident, Yoshiharu Kanno, a father of two children from Watari, noted that "the current trend in Fukushima is that the government is pressing hard for reconstruction and because of this it is becoming 'taboo' to express the health effects of radiation" (World Network for Saving Children from Radiation 2013).

Even after official decontamination began in the Watari district in January 2012, comprising a total of 727 houses, with priority to households with children, residents continued to express their concerns.

> They conducted decontamination once in Watari districts but it was very simple. They lifted the gratings of a road in front of Watari elementary school and took some of the soil and that's it. They did not even use a high-pressure water sprayer. Then, they left bags with the soil for two to three days, so the air contamination rose during that time. Although they used a high-pressure sprayer for the road behind the elementary school, it is well-known that while the sprayer may rinse off radioactive materials, it cannot eliminate them and they just remain somewhere else.
>
> (resident of Watari district, quoted in Human Rights Now 2012: 8)

> The current decontamination work is almost meaningless. I live at the bottom of a hill and there are many houses and apple fields on the hill. There are many side ditches as well. It [water] comes down to side ditches near my

> house. I was told by residents who live on the top of the hill that 'We used a high-pressure sprayer to decontaminate and dumped the water into side ditches. We feel sorry but we had no other choice'. I heard from a person who measured the radiation level of the ditch in early October (2011) that it was 30 to 40μSv/h. If we run the contaminated water down to rice paddies at the bottom would be affected as well. Eventually, the water gets to Abukuma River then to the sea. It might be no problem for people in highlands but it is troublesome for people living in lowlands. Decontamination in one place is not effective, it has to be done in all of the district.
>
> (resident of Watari district, quoted in Human Rights Now 2012: 9)

Radiation "hot spots"

Another concern of residents and NGOs was that, while overall ambient dose levels had decreased in many areas, as indicated in Table 6.1, high levels of contamination remained topically concentrated in "micro hot spots,"[8] typically rain gutters, sewer channels, and fences, even a few meters away from the government-installed monitoring stations. Moreover, although some areas were known to have higher than average levels of radiation and hence needed to be decontaminated, there were hardly any entry restrictions on them. Japan NGO Center for International Cooperation (JANIC) members Toshiyuki Takeuchi and Emiko Fujioka (2013: 5) commented on the different approach to risk communication that resulted after March 2011 in Fukushima in the following way:

> Japan is famous for precautions, posting numerous warning signs everywhere for various risks such as "Beware of bears", "Hazardous volcanic gases eruption. No entry", "Beware of falling rocks", "Slow down" and "Smoking can be hazardous to your health". Then why is only radiation treated differently? There is no warning sign even in the area where radiation levels exceed the legal limit?

Greenpeace Japan also found that contamination had concentrated in many hot-spot places in Fukushima City and argued that these posed serious threats to human health. For example, one year after the disaster, their team found hot spots as high as 70 mSv/hour (at a height of 10 cm) in a parking garage 50 meters from the central train station, and 40 μSv/hour in a water drain next to housing, representing up to 1,000 times normal background levels (Greenpeace International 2014). Later still, at the end of 2014, Greenpeace continued to claim that hot spots existed in Fukushima, citing one in front of a city hospital that measured 1.1 μSv/hour. Although this was one of the highest hot-spot readings at that time, Greenpeace found 70 other places in the city where the amount of radiation recorded exceeded the MOE's long-term target of 0.23 μSv/hour. The MOE disputed Greenpeace's claim that it was underestimating radiation contamination and risks and pointed out that radiation in Fukushima City had steadily diminished over time (Boyd 2014).

What, then, are we to make of these micro hot spots? Since 2011, it has been relatively easy to find locations in Fukushima City that remained above 0.23 μSv/hour, even after decontamination of various areas had taken place. Indeed, apart from the NGOs mentioned above, micro hot spots were discovered in Fukushima City by citizens' groups, local authorities, and government inspectors as they continued their surveys of contaminated areas in 2012 and 2013. Nonetheless, city officials argued that in almost every conceivable case that, even if nothing was done to clean up a place that recorded say over 1 mSv/year, it would eventually decline to 0.5 mSv/year or less after around 30 years, due to natural radioactive decay. In other words, they contended that while the readings certainly appeared high, the newly discovered micro hot spots posed no threat to human health. They believed that total radiation dose was measured not just by the strength of radioactivity in a given area but also by the time a person spent there. Consequently, the small size of hot spots, often in gutters or in stagnant pools, made it all but inconceivable that anyone would receive up to 20 mSv/year, which was the standard used for residential evacuation. Indeed, to reach the higher and certainly more dangerous 100 mSv/year dose would require someone to be continually exposed to the Japanese government's 20 mSv/year limit for 24 hours a day, seven days a week, over a five-year period. Consequently, experts argued that a small amount of radioactive material on a rooftop or in a gutter posed little risk. "People do not normally sleep in one spot in a gutter" (Brumfiel 2011).

Even so, while hot spots might comprise only miniscule amounts of radiation, many residents of Fukushima continued to be nervous, and, as pointed out earlier, "perceived risk" is about more than radiation readings. Indeed, most concerns expressed by residents of Fukushima in the years up to 2014 related to hot spots that had not been decontaminated by municipal authorities. While accepting that radiation levels had declined overall, NGOs such as Greenpeace, for instance, argued that Fukushima residents wanted to know what the radiation effects were in their local neighborhoods, and whether there were build-ups of radiation doses higher than the average recorded across the city (Boyd 2014).

One of the study interviewees, Ms Akemi Shima, a concerned mother of two children, recollected her discussion with a school principal in September 2013, when a hot spot was found measuring 10 μSv/hour adjoining a school zone.

> I asked for the hot spot to be decontaminated, but the school official said: "We only have a few storage areas for contaminated soil and waste. So we won't do it." I countered: "So please secure it, because children use this area." But they did not. Another time I asked the city officials to decontaminate the road crossing where children go, but was told by a city official that as people only spend a few seconds crossing the road then there was no need to make this a priority. But I think that children often spend a whole hour at one spot, and their behavior cannot be regulated.
>
> (interview with Akeshi Shima, Watari, Fukushima, June 2014)

According to the city's website, it had dealt with 283 hot-spot decontamination problems up to mid 2014 (City of Fukushima 2014).

Delays and the storage of waste material

Yet another issue that irked local residents and contributed to their nervousness about low-level radiation was the long delay in completing residential decontamination in certain districts of the city. Apart from the dearth of human resources required to do this work, an additional reason for the delay was that the city administration needed to secure sufficient storage space for radioactive waste. In fact, decontamination activities generated an inordinately large amount of waste material, such as contaminated soil stripped from the ground, together with leaves and twigs, which was bagged in large black plastic containers. Suitable public sites under the control of the municipality or prefecture, such as public yards, urban rubbish dumps, and landfill sites were soon exhausted, leading to the need to build temporary storage sites on private land. The city's website indicated that 9 out of 13 large-scale temporary facilities for waste were either operating on, or planned to operate on, private residential land (City of Fukushima 2014).

Nonetheless, residents opposed the temporary storage of contaminated waste, either on their own or their neighbors' land, due both to the fear of radiation and also uncertainty over how long bags of tainted soil would be left there. Consequently, residential waste in Fukushima generally remained within decontaminated housing lots, either stored above ground on the property or buried with markings for up to three years. City officials indicated that 80 percent of residents had chosen to bury waste on their property (interview with Mr Yudai Ouchi, Assemblyman, City of Fukushima, June 2014). Still, the longer-term intent was to move waste from these locations to a national government facility to be built within the Special Decontamination Area. In mid 2014, agreement was reached in principle with the then Fukushima governor, Yuhei Sato, to build large region-wide temporary storage facilities in Okuma and Futaba in return for 301 billion yen (about 3 billion US dollars) in subsidies (Aoki 2014).[10]

Conclusions

The Japanese government introduced policies and practices to decontaminate low-level radiation in 2011 to protect the health and safety of affected residents, particularly evacuees. But as implied by the two quotes that began this chapter, residents and their NGO supporters have censured decontamination programs for not reducing their anxiety and for not addressing their perception that there was a high risk involved even in low-level radiation, especially for children. As we have seen, the "government approach" was based upon scientific analysis. It contended that while there was no absolutely safe level of radiation, the ICRP reference level of 20 mSv/year was an appropriate threshold for ordering evacuation, and an "acceptable risk" that people in non-evacuation zones would not be exposed to unreasonable hazards: in other words, it was "safe enough." The so-called "citizens' level," the much lower 1 mSv/year, was adopted by the government as a long-term goal for their decontamination program. But as seen in the case of Fukushima City, decontaminating houses and public facilities to this standard proved to be very difficult and costly, even with a substantial budget provided by the national government. Consequently, municipal efforts were not "safe enough" in the eyes of many citizens. Moreover, the hot spots that irked both residents and NGOs were seen by the city as "no great risk." While the city had indeed cleaned up several hot spots, resources and lack of facilities to store all the waste gathered from decontamination meant that they could not decontaminate hot spots in all cases. Indeed, with the knowledge that long-term-radiation levels had declined mainly due to natural weathering effects, it is not hard to conclude that the government's decontamination program appeared to be an "optical solution" to show the public that at least token efforts were being made.

Whether this is true or not, the fieldwork conducted in Fukushima City shows clear three findings. First, local governments in the Intensive Contamination Survey Areas had no recourse but to follow the national standards and procedures set down in the national guidelines. For instance, the city's own Furusato Decontamination Plan essentially duplicated national guidelines—it had to in order for the city to obtain national funding. Second, while the city had a relatively open approach to communicating its decontamination operations, local residents' and NGOs' lack of trust in the Japanese government and TEPCO could not be easily regained. Consequently, local governments in Fukushima appeared sullied as allies of the so-called "nuclear village,"[11] making faith in the decontamination program difficult, especially the technical procedures associated with measuring radiation, and problematizing cooperation with the city over the acceptance of storage areas on residential land. Third, in terms of the "dialogue model" of Palenchar and Heath (2006), there is some evidence that Fukushima City attempted to engage with local residents on their priorities and plans by contacting leaders of neighborhood associations and property owners. However, it appears that there was little or no attempt at collaboration to discuss the appropriate risks of low-level radiation, which some argued was necessary (Fukushima Action Research on Effective Decontamination Operation 2013; Safecast 2013).[12]

Finally, it appears that the "cavernous disconnect" between policy makers and experts on the one hand and residents on the other may widen rather than contract in the near future. Thus, in mid 2014, the national government was reported to be raising the official radiation target level in Intensive Contamination Survey Areas from the current 0.23 μSv/hour to 0.4–0.6 μSv/hour, and moving away from an aerial-radiation basis for decontamination to an individual-exposure basis. The MOE considered that the additional exposure in the new target levels, beyond background levels, would still keep exposures under 1 mSv/year. Under the new policy, the government would determine decontamination needs by using radiation-exposure data collected from personal dosimeters handed out to all residents, rather than rely on airborne monitoring equipment. This change was ostensibly due to empirical data collected from the so-called "glass badge" program, whereby small personal dosimeters recorded cumulative radiation exposure by residents in Fukushima, Date, and Soma cities, which found that the vast majority had not exceeded 1 mSv over one year, rendering the assumptions of the MOE model described earlier obsolete and invalid. While local municipalities believed that this approach would be a more realistic target for decontamination programs, doing away with ineffective decontamination work, local residents perceived the planned changes as all about "cost-performance," as the original and lower 0.23 μSv/hour was seen as too costly and time-consuming (see also Chapter 7). Some municipalities welcomed the move, saying it would allow them to scale down decontamination efforts in areas where radiation levels were unlikely to go down significantly. Others, however, worried that residents would be merely confused (Fukushima Minpo 2014).[13]

Acknowledgments

The author would like to acknowledge the many interviewees in Fukushima and other parts of Japan that contributed their time. In particular, I would like to thank the assistance of Professor William McMichael and Professor Nori Fujimoto, both of Fukushima University, and Mr. Yuta Hirai, freelance journalist. Eric Leinberger drew the figures. This study was supported by a Canadian Social Science and Humanities Research Council (SSHRC) Grant #435-2014-0700.

Notes

1 A caveat should be mentioned here, as there are no reliable numbers regarding the prevalence of citizens anxious about low-level radiation in Fukushima. My own observation is that after 2013, as there was more information about radiation in the community, together with a gradual fall in radiation levels, fewer people expressed concern. However, just because problems were not mentioned does not mean that people no longer feared radiation. For the view of mothers who were interviewed in Fukushima, see Ash (2012), and for mothers from Fukushima who had evacuated with their children to Tokyo, see the work of Slater *et al.* (2014). Ikeda (2014) also discusses how the risk of radiation has been constructed in Fukushima.

2 Interviews conducted in 2013 and 2014 in Fukushima City included meetings with one former mayor, an assemblyman, three academics, two journalists, municipal and prefectural officers, and a concerned mother.

3 A sievert (Sv) is a unit referring to the extent of radiation damage. It is also used to represent the risk of the effect of external radiation from sources outside the body. One thousandth of 1 Sv is 1 mSv (millisievert), and one thousandth of 1 mSv is 1 μSv (micro-sievert) (Lombardi 2006).

4 Scientists know how high-dose exposure to radiation affects the risk of cancer but do not have exact information on low doses. Therefore, they must rely on mathematical models to predict the likely effects of low exposure. The Linear no-threshold model (LNT) assumes that the adverse effects of radiation (including low-level radiation) are proportional to the radiation dose, all the way down to near-zero dose levels. Whether the LNT model describes the reality for small-dose exposures is disputed (Meister 2005).

5 Interestingly, the construction companies that won contracts to decontaminate Fukushima were the very same ones involved in building the stricken Daiichi NPP in the 1960s (Aoyama 2012). The budget for the government's decontamination program,

including long-term storage of waste, was estimated at 1.9 trillion yen (about 19 billion US dollars) in April 2014 (Ministry of Environment 2014).

6 The airborne standards dealt with in this study concern people's external contamination exclusively, not internal contamination, particularly from food, which warrants its own separate study of exposure risks (Konkel 2014).

7 Radiation released from the Fukushima Daiichi NPP accident involved roughly half cesium-134 and half cesium-137. The biological half-life of cesium-134 is around two years, and that of cesium-137 is about 30 years (Takahashi 2014).

8 The term "micro hot spot" is used to distinguish these small-scale areas from the hot-spot neighborhoods within the city.

9 As indicated earlier, a dose reading of 20 mSv, the lowest part of the 20–100mSv limit range recommended by international organizations, was used as a threshold to require evacuation. For cities outside the mandatory evacuation zones, the government under certain circumstances (mainly where a 20 mSv/year level or higher could be ascertained) established "Specific Spots Recommended for Evacuation" to support any residents who wished to evacuate. Importantly, compensation was available in these cases for residents who decided to do so.

10 At municipal elections held in Fukushima Prefecture during the end of 2013, incumbent mayors in Fukushima City, Koriyama City, Iwaki City, and the town of Tomioka were ousted from office due to residents' frustration over delays in decontamination. At the time, in Fukushima City, about 115,000 houses were targeted for decontamination work but only 18 percent had been completed (*Asahi Shimbun* 2013). As for the national government's own decontamination program, work had not even started in parts of the 11 municipalities of the Special Decontamination Areas closer to the NPP (Aoki 2014).

11 The "nuclear village" is a term that describes the groups who benefit from nuclear power. For example, the Ministry of Economy, Trade and Industry, nuclear-reactor manufacturers, electric-power companies, electric-power associations, local governments, and so on. This term has a cynical connotation and is used mostly by opponents of nuclear power (Vivoda 2014).

12 Safecast (2013) has underscored how important it is for disaster-hit areas to set up opportunities for residents, community groups, trusted medical institutions, and individuals who are trusted by local residents to share trustworthy information. Unfortunately, administrative support for these types of roundtable is currently limited.

13 Keizo Ishii, director of the Research Center for Remediation Engineering of Living Environments Contaminated with Radioisotopes, Tohoku University, cautioned:

> Many residents of Fukushima have deliberately stayed indoors since the nuclear disaster. If they start to go out like they used to before the quake, then individual radiation doses might go up and will not necessarily fall below the 1 millisievert threshold. As such, we should aim for continued use of aerial figures for decontamination.
>
> (Fukushima Minpo 2014)

References

Allison, Wade. 2009. *Radiation and Reason: The Impact of Science on a Culture of Fear*, New York: Wade Allison Publishing.

Aoki, Mizuho. 2014. "Opposition to Waste Storage Complicates Project: Fukushima Cleanup Going Painfully Slow." *Japan Times*, September 22. Accessed September 25, 2014. www.japantimes.co.jp/news/2014/09/22/reference/fukushima-cleanup-going-painfully-slow/#.VVV0lzlVu0s.

Aoyama, Teiichi. 2012. Decontaminate the Fukushima Decontamination Project: "Josen" (the decontamination of radioactive substances) equals to "Isen" (the relocation of contaminated materials) and to the "Concession" for the related organizations. Accessed January 19, 2013. www.eforum.jp/Decontaminate%20the%20Fukushima%20decontamination%20project.pdf.

Asahi Shimbun. 2013. "Frustrated Voters Dump Incumbent Mayor in Fukushima Election," November 18. Accessed November 21, 2013. http://ajw.asahi.com/article/0311disaster/fukushima/AJ201311180075.

Ash, Ian T. 2012. "More Mothers and Their Children." Documenting Ian, blog, December 19. Accessed March 20, 2013. http://ianthomasash.blogspot.ca/2012/12/more-mothers-and-their-children.html.

Boyd, John. 2014. "Study: Fukushima Health Risks Underestimated." *Aljazeera*, November 15. Accessed November 11, 2013. www.aljazeera.com/indepth/features/2014/11/will-japan-reopen-nuclear-plants-fukushima-20141111112653560643.html.

Brasnor, Philip. 2012. "Local Media Are Too Vague on Fukushima Radiation Risk." *Japan Times*, November 11. Accessed January 19, 2013. www.japantimes.co.jp/news/2012/11/11/national/media-national/local-media-are-too-vague-on-fukushima-radiation-risk/#.VVO_qjlVu0s.

Brumfiel, Geoff. 2011. "Fukushima 'Hot Spots' Raise Radiation Fears." *Nature*, October 14. Accessed January 19, 2013. www.nature.com/news/2011/111014/full/news.2011.593.html.

Brumfiel, Geoff and Ichiko Fuyuno. 2012. "Japan's Nuclear Crisis: Fukushima's Legacy of Fear." *Nature*, 483(7388): 138–40.

City of Fukushima. 2012. *The Implementation of Decontamination Plan of Fukushima*. 2nd ed. Accessed June 9, 2014. www.city.fukushima.fukushima.jp/uploaded/attachment/10812.pdf. [In Japanese.]

City of Fukushima. 2014. "Fukushima Decontamination Information Centre." Accessed June 9, 2014. www.city.fukushima.fukushima.jp/soshiki/76/jyosen-cente.html. [In Japanese.]

Embassy Science Fellow Team. 2013. *Report of the United States Embassy Science Fellows Support to the Government of Japan – Ministry of the Environment*. Accessed December 30, 2014. http://josen.env.go.jp/en/documents/pdf/workshop_july_17-18_2013_04.pdf.

Fischhoff, Baruch, and John Kadvany. 2011. Risk: A Very Short Introduction, Oxford: Oxford University Press.

Fischhoff, Baruch, Paul Slovic, Sarah Lichtenstein, Stephen Read, and Barbara Combs. 2000. "How Safe is Safe Enough? A Psychometric Study of Attitudes Towards Technological Risks and Benefits." In *The Perception of Risk*, edited by Paul Slovic, 80–103. New York: Taylor and Francis.

Friends of the Earth Japan. 2013. "Fukushima Poka-Poka Project Activity Report, Friends of the Earth Japan." Accessed January 15, 2014. www.foejapan.org/en/news/130301.html.

Fukushima Action Research on Effective Decontamination Operation. 2013. *Challenges of Decontamination, Community Regeneration and Livelihood Rehabilitation*. Kanagawa, Japan: Institute for Global Environmental Strategies. Accessed January 15, 2014. pub.iges.or.jp/modules/envirolib/upload/4718/attach/web_FAIRDO_2nd_Discussion_Paper_E_130906.pdf.

Fukushima Minpo. 2014. "New Radiation Measurement Method Spreads Confusion." *Japan Times*, July 20. www.japantimes.co.jp/news/2014/07/20/national/new-radiation-measurement-method-spreads-confusion/#.VVPB0DlVu0s.

Greenpeace International. 2014. "Radiation Surveys, Fukushima, 30 October." Accessed October 2, 2014. www.greenpeace.org/international/en/campaigns/nuclear/safety/accidents/Fukushima-nuclear-disaster/Radiation-field-team/.

Hanai, Toru and Elaine Lies. 2014. "The Children of Japan's Fukushima Battle an Invisible Enemy." Reuters, March 11. Accessed June 20, 2014. www.reuters.com/article/2014/03/10/us-japan-fukushima-children-idUSBREA280RJ20140310.

Hirakawa, Hideyuki and Masashi Shirabe. 2015. "Rhetorical Marginalization of Science and Democracy: Politics of Risk Discourse on Radioactive Risks in Japan." In *Lessons from Fukushima*, edited by Yuko Fujigaki, 57–86. Heidelberg: Springer International Publishing.

Human Rights Now. 2012. *Investigative Report on Fukushima City and Koriyama City: Fact-finding Mission Conducted on November 26 and 27, 2011*, Tokyo: HRN. Accessed January 19, 2013. http://hrn.or.jp/eng/activity/Investigative%20Report%20on%20Fukushima%20City.pdf.

Ikeda, Yoko. 2014. "The Construction of Risk and the Resilience of Fukushima in the Aftermath of the Nuclear Power Plant Accident." In *Japan Copes With Calamity: Ethnographies of the Earthquake, Tsunami and Nuclear Disasters of March 2011*, edited by Tom Gill, Brigitte Steger, and David H. Slater, 151–76. Bern: Peter Lang.

Konkel, Lindsey. 2014. "Data for Disaster Planning: Risk Factors for Internal Radiation Exposures after Fukushima." *Environmental Health Perspectives*, 122(6): 166.

Lombardi, Max H. 2006. *Radiation Safety in Nuclear Medicine*. 2nd ed, Boca Raton, FL: CRC Press.

Mahr, Krista. 2011. "In Fukushima City, Decontamination Begins. But What to Do with the Radioactive Waste?" *Time*, August 9. Accessed January 15, 2014. http://science.time.com/2011/08/09/in-fukushima-city-decontamination-begins-but-what-to-do-with-the-radioactive-waste/.

Meister, Kathleen. 2005. *The Health Effects of Low-Level Radiation*, New York: American Council on Science and Health. Accessed January 15, 2014. http://acsh.org/2005/09/the-health-effects-of-low-level-radiation/.

Ministry of Environment. 2013a. *Decontamination Guidelines*, 3rd ed., Tokyo: Ministry of Environment.

—— 2013b. *Progress on Off-site Cleanup Efforts in Japan, July 17*, Tokyo: Ministry of Environment. Accessed June 9, 2014. http://josen.env.go.jp/en/pdf/progressseet_progress_on_cleanup_efforts.pdf?141022.

—— 2014. *Off-site Decontamination Measures*, Tokyo: Ministry of Environment. Accessed June 9, 2014. http://josen.env.go.jp/en/#top07.

Office of the Deputy Chief Cabinet Secretary. 2011. Cabinet Report of the Working Group on Risk Management of Low-dose Radiation Exposure, December 22, 2011, Office of the Deputy Chief Cabinet Secretary. www.cas.go.jp/jp/genpatsujiko/info/twg/Working_Group_Report.pdf.

Onishi, Norimitsu and Martin Fackler. 2011. "Japan Held Nuclear Data, Leaving Evacuees in Peril." *New York Times*, August 8. Accessed July 15, 2013. www.nytimes.com/2011/08/09/world/asia/09japan.html?_r=0.

Palenchar, Michael J. and Robert L. Heath. 2006. "Responsible Advocacy through Strategic Risk Communication." In *Ethics in Public Relations: Responsible Advocacy*, edited by Kathy Fitzpatrick and Carolyn Bronstein, 131–54. Thousand Oaks, CA: Sage Publications.

Renn, Ortwin. 2009. "Risk Communication: Insights and Requirements for Designing Successful Communication Programs on Health and Environmental Hazards." In

Handbook of Risk and Crisis Communication, edited by Robert L. Heath and H. Dan O'Hair, 80–98. New York: Routledge.

Safecast. 2013. "Decon or Con? How Is Remediation Being Managed and How Effective Is It?" Safecast, August 17. Accessed January 19, 2015. http://blog.safecast.org/2013/08/decon-or-con-how-is-remediation-being-managed-and-how-effective-is-it/.

Sanada, Yukihisa, Takeshi Sugita, Yukiyasu Nishizawa, Atsuya Kondo, and Tatsuo Torii. 2014. "The Aerial Radiation Monitoring in Japan After the Fukushima Daiichi Nuclear Power Plant Accident." *Progress in Nuclear Science and Technology*, 4: 76–80.

Slater, David H., Rika Morioka, and Haruka Danzuka. 2014. "Micro-Politics of Radiation." *Critical Asian Studies*, 46(3): 485–508.

Slovic, Paul. 2000a. "Perception of Risk." In *The Perception of Risk*, edited by Paul Slovic, 200–31. Abingdon: Routledge.

—— 2000b. "Perception of Risk from Radiation." In *The Perception of Risk*, edited by Paul Slovic, 264–74. Abingdon: Routledge.

—— 2000c. "Trust, Emotion, Sex, Politics and Science: Surveying the Risk-assessment Battlefield." In *The Perception of Risk*, edited by Paul Slovic, 390–412. Abingdon: Routledge.

Slovic, Paul, James H. Flynn and Mark Layman. 2000. "Perceived Risk, Trust and the Politics of Nuclear Waste." In *The Perception of Risk*, edited by Paul Slovic, 275–84. Abingdon: Routledge.

Streffer, Christian, H. Bolt, D. Follesdal, P. Hall, J.G. Hengstler, D. Oughton, and E. Rehbinder. 2004. *Low Dose Exposures in the Environment: Dose-effect Relations and Risk Evaluation*. Berlin: Springer-Verlag.

Suzuki, Itoko and Yuko Kaneko. 2013. *Japan's Disaster Governance: How Was the 3.11 Crisis Managed?* New York: Springer.

Takahashi, Sentaro, ed. 2014. *Radiation Monitoring and Dose Estimation of the Fukushima Nuclear Accident*, Tokyo: Springer.

Takeuchi, Toshiyuki and Emiko Fujioka. 2013. "Case Story: The Reality of the Radioactive Contamination at Fukushima: Was It Really the Environment That Was Contaminated, or Was It the Heart of the People?" *Stories and Facts from Fukushima*, June 5. Accessed July 15, 2013. http://fukushimaontheglobe.com/wp-content/uploads/Stories-Facts-from-Fukushima-2.pdf.

Vivoda, Vlado. 2014. *Energy Security in Japan: Challenges after Fukushima*, Farnham: Ashgate Publishing.

Wilson, Richard and Edmund A. C. Crouch. 2001. *Risk–benefit Analysis*, Cambridge, MA: Harvard University Press.

World Network for Saving Children from Radiation. 2013. "Why I Myself Cannot Evacuate, a Father in Fukushima Confesses." Accessed June 18, 2014. www.save-children-from-radiation.org/2013/02/21/why-i-myself-cannot-evacuate-a-father-in-fukushima-confesses/.

Wrixon, Anthony D. 2008. "New ICRP Recommendations." *Journal of Radiological Protection*, 28(2): 161–8.

7 Decontamination-intensive reconstruction policy in Fukushima under governmental budget constraint

Noritsugu Fujimoto

Introduction

The Great East Japan Earthquake struck the predominantly rural Tohoku region the hardest, and the subsequent tsunami devastated over 500 km of its Pacific coast. The number of deaths and injuries caused by the earthquake and the tsunami were modest compared to those resulting from similar geophysical events elsewhere in the world (Guha-Sapir *et al.* 2014). Nevertheless, the recent Japanese disaster produced one of the largest natural-disaster-induced economic losses in modern history (even without including politically contested damages from the nuclear accident as described in this chapter), ahead of the Hanshin-Awaji Earthquake in 1995, which hit heavily urbanized areas of the country. What made the Great East Japan Earthquake disaster so costly, of course, was the severe accident at the Fukushima Daiichi Nuclear Power Plant (NPP) of the Tokyo Electric Power Company (TEPCO) (Rabl and Rabl 2013). A series of hydrogen explosions of the reactors resulted in the release of radioactive materials in extensive areas in eastern Japan, affecting most significantly the eastern part of Fukushima Prefecture and its adjacent areas. Radioactive contamination does not easily fit conventional categories of natural disasters (Guha-Sapir *et al.* 2012). In particular, the locations and populations affected by radioactive contamination are not easily determined by conventional guidelines for analyzing natural disasters, yet such information is essential in calculating economic damages and appropriate compensation. Consequently, determining the extent and magnitude of radioactive contamination has become a highly complex and politically charged issue in post-3.11 Japan.

After the nuclear accident, 8,000 monitoring posts were set up throughout eastern Japan by the Nuclear Regulation Authority (NRA) to monitor the spread of radioactive material (in terms of air dose) to help determine the locations and extent of evacuation. For example, 394 monitoring posts were allocated to Fukushima City, with an area of 767 km^2, which means that there is one post for approximately every 2 km^2. With these measurements and 20 mSv/year as the threshold radiation level, the national Cabinet Office ordered persons living within 20 km of the plant and in other "hotspots" in several small towns and villages to evacuate from their homes soon after the nuclear accident. Residents of these areas became entitled to receive financial compensation from TEPCO—not directly from the local or central government, because the company is liable for damages under the Atomic Energy Damage Compensation Law in Japan. However, the government does use public funds to keep TEPCO afloat through loans and other means. By doing so, it can keep the private company directly legally responsible for the accident while still supporting the company and keeping the country's energy policy intact (Aldrich 2011).

Even within the current scheme, the government's practice is problematic on at least two counts. First, the current monitoring system, with a 2 km grid, provides far too sparse area coverage, given that contamination can vary considerably at much smaller resolutions (Chapters 6 and 8). Radioactivity measurements by some non-profit organizations, non-governmental organizations, and universities have revealed the presence of many radioactive hotspots outside the 20 km evacuation zones (Citizen's Radioactivity Measuring Station 2013). Second, the current safety limit of 20 mSv/year is arguably too high a threshold for evacuation. Five mSv/year is typically the maximum level of radiation allowed in "radiation controlled areas" such as medical X-ray rooms sealed with "NO ENTRY" signs in clinics and hospitals. In the Chernobyl accident, 5 mSv/year was used by the government to determine evacuation zones (Lych and Pateeva 1999). If we apply the Chernobyl standard to Fukushima, two thirds of Fukushima Prefecture, including much of the Hama-Dori (coastal area) and the Naka-Dori (central area), would be included within the evacuation zones. According to the newly established Reconstruction Agency, there are about 50,000 voluntary evacuees, many of whom used to live in these "would-be" evacuation areas, and who appear to sense high health risks even outside the official boundaries. These "volunteer evacuees" do not receive compensation from TEPCO.

This chapter investigates the political economy of the nuclear-accident evacuation zones. I first review major explanations for why the government has set the relatively limited current evacuation zones and why many residents still live in areas that have above-normal radiation levels. I will then argue that TEPCO and the central government, by underplaying the real magnitude of the contamination and damage, are seeking to work within national budgetary constraints. The chapter finally demonstrates how the decontamination of radiation-contaminated land is used as a fiscally cheaper and politically easier alternative to larger-scale evacuation and examines what such policy implies for the regional economies of Fukushima.

Political economy of nuclear-incident evacuation zones

Conventional explanations

Why has the central government excluded central Fukushima (Naka-Dori), which includes the prefectural capital Fukushima City, from the evacuation zone, even though much of it would be included by conventional standards; and why do many people continue to live there? Let me first review the major explanations, or what I would call "myths," for the exclusion of central Fukushima from evacuation zones before explaining the dynamics of zoning from the viewpoint of national budgetary constraints.

The first common belief is that land is scarce in Japan. In particular, it is assumed that there is no room to build large, new cities outside of central Fukushima, where most of the population of Fukushima Prefecture is concentrated, with 213 persons/km^2. Such a belief is clearly unfounded because excess supply is always observed in Japan's property market. In fact, as part of the ongoing discussion of *Sento* (national capital relocation), several local governments in Fukushima and Miyagi Prefectures have long sought to become the nation's capital by hosting the Diet, Bureaucracy, and Supreme Court and by providing space for 600,000 new residents. Furthermore, the population density of Fukushima Prefecture is actually one of the lowest in the country, with only 140 persons/km^2. Therefore, at least from an aggregate perspective, there is ample land on which to relocate people from areas with contamination above 5 mSv/year.

The second often heard explanation is that residents of Fukushima remain where they are to protect their ancestral lands, passed down through generations. This is certainly not true, because most current landowners in Japan received their land after World War II. In 1947, the Agrarian Land Reform promoted by General Headquarters of the Allied Forces transferred ownership of land from 2.3 million landlords to tenant farmers. Even though Fukushima's economy was highly dependent on agriculture at that time, only a small fraction of current landownership dates back to the pre-war period.

The third explanation relates to transportation issues. Some people seem to believe that the Tohoku Shinkansen (bullet trains) and the Tohoku Expressway, connecting Tokyo and major cities in Tohoku, make central Fukushima a strategically important corridor. In this view, if these high-speed transport systems come to halt, the economic loss due to crippled supply chains in northern Japan will be tremendous. We must keep in mind, however, that these high-speed transport systems are not as intensively used as many assume. The frequency of the Tohoku Shinkansen is much lower than that of the Tokaido–Sanyo Shinkansen, which connects cities such Tokyo, Nagoya, Osaka, Hiroshima, and Kitakyushu. The bullet trains from Tokyo to Fukushima run only every half an hour. Even if rapid trains from Tokyo to Sendai, which skip Fukushima and Koriyama stations, are included, the frequency is only every 20 minutes. In contrast, the frequency of Tokaido (Tokyo–Osaka) Shinkansen is every 5 minutes and the Sanyo (Osaka–Fukuoka) Shinkansen every 10 minutes. Similarly, the Tohoku Expressway is not as intensively used as other major expressways, such as the Tomei Expressway (Tokyo–Nagoya), Meishin (Nagoya–Kobe), and Sanyo (Kobe–Shimonoseki). Furthermore, even if the current transport systems in central Fukushima were to come to a halt, detour and bypass routes can be opened relatively quickly. During the Hanshin-Awaji Earthquake in 1995, which paralyzed key high-speed transportation systems around Kobe City, bypass routes, including air travel, quickly opened up and provided alternative connections. The share of air transportation to Kobe from the west (from cities such as Okayama, Hiroshima, Yamaguchi, and

Fukuoka) increased rapidly, permanently replacing some of the share of the Sanyo Shinkansen.

The final widespread belief concerns differences in political economic systems, contrasting centrally planned and democratic market economies. The former system, such as that of the Soviet Union, the argument goes, had the power to require residents to evacuate from contaminated zones around Chernobyl. On the other hand, in Japan, supposedly a democratic market economy, the government cannot force residents to leave. Differences in emergency response and reconstruction policy between Chernobyl and Fukushima are indeed worth studying (Voloshin 1999). Nevertheless, most, if not all, countries embody both planned and market economies to varying degrees. The Japanese Cabinet Office authorized and essentially enforced the evacuation zone within a 30 km radius of the Fukushima Daiichi plant soon after the incident, indicating that the state indeed has the power to relocate its citizens if it so determines.

Current evacuation zones

The Prime Minister and his Cabinet issued the evacuation order after the explosion of the reactors at the Fukushima Daiichi plant, but nobody seems to know who is actually responsible for drawing the specific evacuation zones. There is no rigorous scientific rationale determining that people within a 20 km radius around the plant must leave and that contaminated areas with below 20 mSv/year need not be designated evacuation zones. For example, epidemiologic research on the effects of radiation exposure is based almost entirely on statistical inferences; hence, it is not possible to determine the precise amount of radiation (in mSv/year) that is dangerous to humans (Citizen's Radioactivity Measuring Station 2013). Empirical epidemiological research on the actual health effects of high levels of radiation is limited to very few cases from Chernobyl, Hiroshima, and Nagasaki. Therefore, there is weak epidemiological evidence to suggest that areas with radiation levels between 5–20 mSv/year are safe and can be excluded from compulsory evacuation.

Furthermore, even if we tentatively accept 20 mSv/year as a legitimate threshold, actual radioactive contaminated areas do not always coincide with the current evacuation zones. Central Fukushima, with two medium-sized cities and several small cities, was never authorized as an evacuation zone despite the relatively high radioactivity with more than 20 mSv/year in some parts of the region. Moreover, if we change the spatial resolution of the analysis, there are numerous hotspots of radioactive contamination that cannot be detected by the current 2 km-grid monitoring system.

What would happen if, instead of accepting the current 20 mSv/year standard, we applied the Chernobyl standard of evacuation zones (i.e. 5 mSv/year threshold)? Large sections of central Fukushima would be added to the current evacuation zones, and an estimated 1.15 million people would have to be evacuated (Fujimoto 2015a). In this scenario (Scenario 2 in Table 7.1), an astronomical additional amount of compensation must be paid by the government, TEPCO, or both. Under the guidelines designed by the Nuclear Damage Liability Facilitation

Fund (NDF), 3.6 trillion yen for four years of "mental damage" has been paid to the 88,000 evacuees who used to live within the 20 km radius of the plant. This money does not even include all the lost real-estate values, lost business profits, and health damages experienced by evacuees (TEPCO 2013).

What these boil down to is that the economic damage of the Fukushima nuclear accident is too heavy for the government and TEPCO to cover if central Fukushima is defined as "damaged" and included in official evacuation zones. Before the nuclear accident, the number of residents around the Fukushima Daiichi plant was one of the smallest among all nuclear-hosting regions in Japan (Figure 7.1). Only

Table 7.1 Three scenarios of evacuation and compensation

	Scenario 1 (In force)	*Scenario 2 (Chernobyl standard)*	*Scenario 3 (Pre-disaster standard)*
Radiation standard	20 mSV/year	5 mSV/year	1 mSV/year
Evacuation zone	20 km radius	Approx. 80 km radius	No specific circle drawn
Evacuees (persons)	88,000	1,150,000	3,200,000
Compensation (trillion yen)	0.9	12.8	35.6
Impact on the national economy	Little influence	Critical effects on the national budget and economy	Governmental bankruptcy

Source: Author, based on various sources.

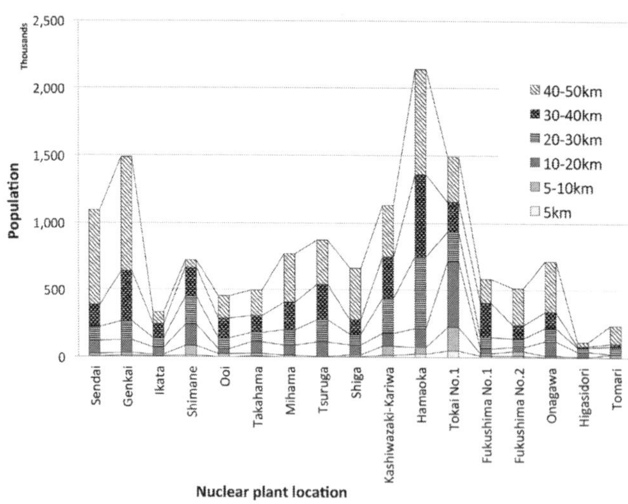

Figure 7.1 Population density around nuclear plants in Japan, 2010.

Source: Modified by the author, based on Nuclear Regulation Authority (2006).

74,000 people lived within a 20 km radius of the plant. What has become clear after this disaster is that no matter how small the resident population around the plant is, an enormous amount of compensation will become necessary.

Macro balance and limits of the national budget

As mentioned above, under the current compensation scheme, it is TEPCO, not the central government, that pays financial compensation for damages from the nuclear accident. Nevertheless, if central Fukushima were included in the evacuation zone, not only might TEPCO go bankrupt, but the central government would also face a large increase in its fiscal deficit. Because the supervisory responsibility of nuclear-plant installations, as public utilities, belongs to NRA, a government agency, the government would be the natural guarantor of TEPCO if it were to fail. This also incentivizes the central government to keep central Fukushima out of official evacuation zones and to keep TEPCO afloat, else the Japanese national economy could face a serious crisis from the enormous sums required for compensation.

Japan's economy has so far managed to escape paralysis, but its public finances rapidly deteriorated during the 2000s due to increased spending on social security and public investment along with shrinking tax revenues (Cabinet Office 2014). Total tax revenues were below the amount of government bond issues for five consecutive years from 2009. The national government debt as a percentage of tax and all other revenues exceeded 104 percent in 2012.[1] If central Fukushima is included in the evacuation zones, at least 800,000 people will be entitled to compensation in addition to the existing 88,000. Even if they are only paid for "mental damage," the additional compensation costs for these 800,000 people could reach 6 trillion yen per year, six times as much as the current costs for those who used to live within the 20 km-radius evacuation zones. This amount nearly equals the annual sales of TEPCO in 2012, 5.9 trillion yen, clearly more than the company alone could bear, and it would have to be supported by the state in one way or another. Yet the government would find it almost impossible to provide such compensation within its current budget constraints.

In theory it is possible for the Bank of Japan to directly underwrite government bonds on the pledge of the central government to impose a massive tax increase, but those actions would bring on hyperinflation (Kobayashi 1975) and redistribution of national wealth between regions. In other words, compensating the residents of central Fukushima through fiscal policies would cause a sudden and profound structural change in the economy. Current policy makers in Japan do not have the courage, or strong incentives, to undertake an action that would be as drastic as the Meiji Restoration (1868) or the Agrarian Land Reform (1947). This is another reason why central Fukushima has never been, and is unlikely to be, designated an evacuation zone.

For comparison, if the same kind of accident were to happen at other nuclear power plants in more urban settings, such as the Tokai Daiich, Hamaoka, Shimane, or Kashiwazaki-Kariwa plants, the population within a 30 km radius would easily exceed 400,000 in each case (Figure 7.2). No electric-power company could bear

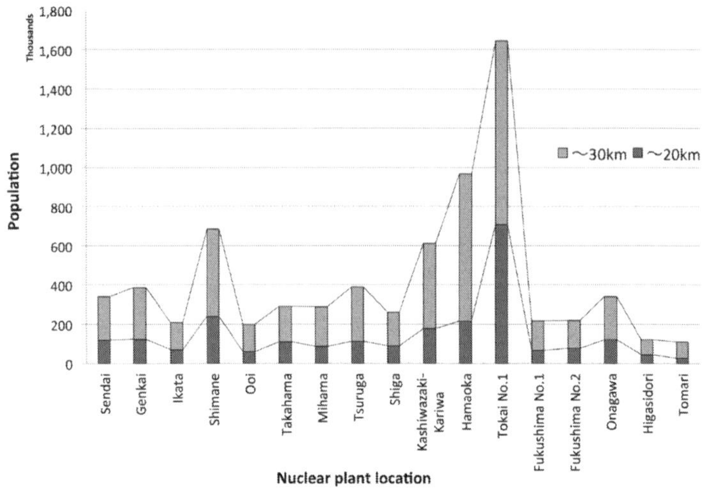

Figure 7.2 Number of persons to be compensated in case of nuclear accident equivalent to the Fukushima nuclear accident.

Source: Modified by the author, based on Nuclear Regulation Authority (2006).

an expense for such compensation, which would easily exceed its annual revenue (Elliott 2013). If we follow the budgetary-limit explanations, it is conceivable that only 5 km-radius areas would be designated as evacuation zones if these four power plants had an accident similar to the Fukushima Daiichi accident. On the other hand, if a similar accident were to happen in several lower-population-density zones, such as Higashidori or Tomari, larger evacuation zones, perhaps over 30 km radius, could be designated for evacuation because their smaller populations would require relatively limited compensation.

Decontamination-intensive reconstruction policy: industrial structure of Fukushima before and after 3.11

In place of more extensive compulsory-evacuation areas, the governmental-industrial complex in post-3.11 Japan has promoted what I call a "decontamination-intensive reconstruction policy." Under this policy, large portions of the reconstruction budget of the central and local governments are spent on decontamination projects, which are typically contracted to general construction companies. In essence, this policy preserves the structure of orthodox public investment in projects such as dams, roads, and ports (Guha-Sapir *et al.* 2013) that has dominated the political economic system of postwar Japan, and which is often referred to as developmentalism (Johnson 1982; Murakami 1996).

Before exploring why decontamination projects are heavily promoted in Fukushima, it is first necessary to understand the changes in the prefectural economic structure. Figure 7.3 shows the economic structure of Fukushima

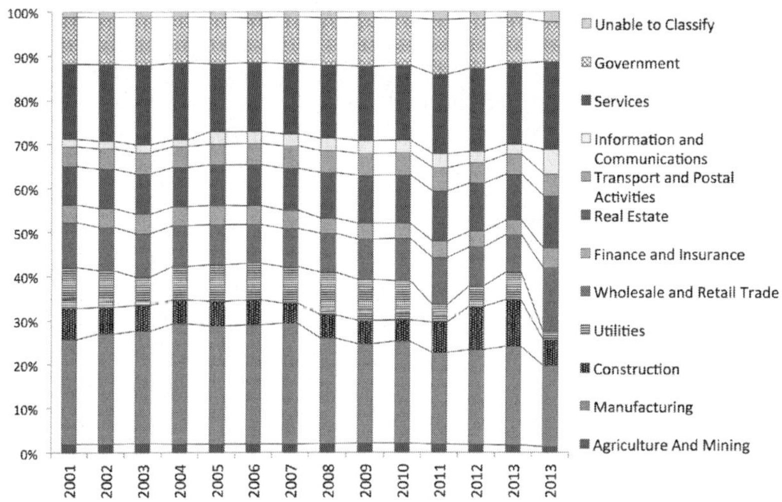

Figure 7.3 Change in economic structure of Fukushima Prefecture in terms of percent of prefectural GDP, 2001–13.

Source: Cabinet Office (2015).

and its change, indicated by the share of each sector's GDP. Fukushima's economic structure, compared to the nation as a whole, is characterized by a high dependence on three sectors: utilities (including electricity), construction, and government. For example, in 2010, utilities accounted for 8.8 percent of total prefectural GDP, which is considerably higher than the average of all prefectures, 2.5 percent, in the same year. This high share reflects the fact that Fukushima had two nuclear plants, Fukushima Daiichi (six reactors) and Daini (four reactors), along the Pacific coast.

Construction's share, which includes civil engineering based on public investment, has also been higher in Fukushima than in the nation as a whole. It accounted for 7.0 percent of prefectural GDP in 2001, which is higher than the average of all prefectures, 6.8 percent, in the same year; however, it had dropped to 4.8 percent by 2010 (Figure 7.4). From the end of World War II until just before the Koizumi administration (2001–6), an enormous amount of government investment in the form of infrastructure construction had been poured into low-income, peripheral regions of the country such as Tohoku, Hokkaido, Hokuriku, and South Kyushu. Fukushima Prefecture, with relatively poor locations for propulsive industries (for example, petrochemical, iron and steel, and automobile industries) in the 1960s and 1970s, was no exception. Many unprofitable projects, such as the Tohoku Shinkansen, small- and medium-sized seaports, airports, and minor highways were built by public investment as part of massive fiscal-transfer policies aiming for "more balanced national development." The Koizumi administration challenged this dominant postwar national-development-policy orientation in the early 2000s under the name of structural reform (Fujimoto 2014).

The Koizumi structural reform resulted in sudden reductions in public investment. The central government drastically cut subsidies to rural prefectures. This reform had a considerable impact on the construction industry nationwide. The decline in the GDP share of the construction industry in Fukushima in the early 2000s reflected this national trend. Nevertheless, there is another factor that is specific to Fukushima that contributed to this decline. In 2006, the governor of Fukushima Prefecture was arrested by the Tokyo District Public Prosecutor's Office for accepting a bribe from a construction company. Consequently, the prefectural government reformed its auction system for local public projects and, in 2007, installed a new competitive-bidding system for all prefectural contract works. This change also contributed to the decline in the output of Fukushima's construction industry from 2006–10 (Figure 7.3).

The declining relative importance of the construction industry reversed after the Great East Japan Disaster in 2011. Its GDP share increased to 7.0 percent in 2011 and 9.5 percent in 2012, doubling its share from 2010. On the other hand, the share of the electric-power industry declined further, falling to 3.8 percent in 2011 and 4.4 percent in 2012, less than half of its share in 2010. It is important to realize, however, that the revival of the construction industry since 2011 does not reflect the rebuilding of the infrastructure destroyed by the earthquake and tsunami, as I discuss below.

A macroeconomic account of the decontamination policy

Decontamination was introduced in Fukushima as a substitute for a broader evacuation in order to maintain the government's fiscal balance as well as the existing political economic structure (Figure 7.4). Decontamination of radiation-contaminated areas essentially involves the relocation of radioactive substances, usually deposited on plants, soil, and dust, from one affected area to another (Munro 2013). For example, workers use pressurized air or water to wash walls and ceilings, or they may physically remove topsoil and transport it to designated areas (see also Chapter 6). Essentially, these are labor-intensive activities with little room for technological innovation. In this way, although decontamination may be considered a public project, its economy-wide productivity-enhancing effects are limited, unlike typical construction projects such as airports or bullet trains.

Money from the National Treasury has been spent on decontamination chiefly by way of the Ministry of Environment and the Reconstruction Agency. The approved disbursement for decontamination was 72 billion yen in 2013 (Table 7.2). This is the second largest budget item after the reconstruction of cities and towns, totaling 166 billion yen, and is followed by a special grant-in-aid (60 billion yen) and the reserve fund for Fukushima (60 billion yen). Decontamination, therefore, is prioritized in the post-disaster reconstruction budget. Nevertheless, it is still far smaller than what it would cost the government and TEPCO to provide the "mental damage" compensation alone, estimated

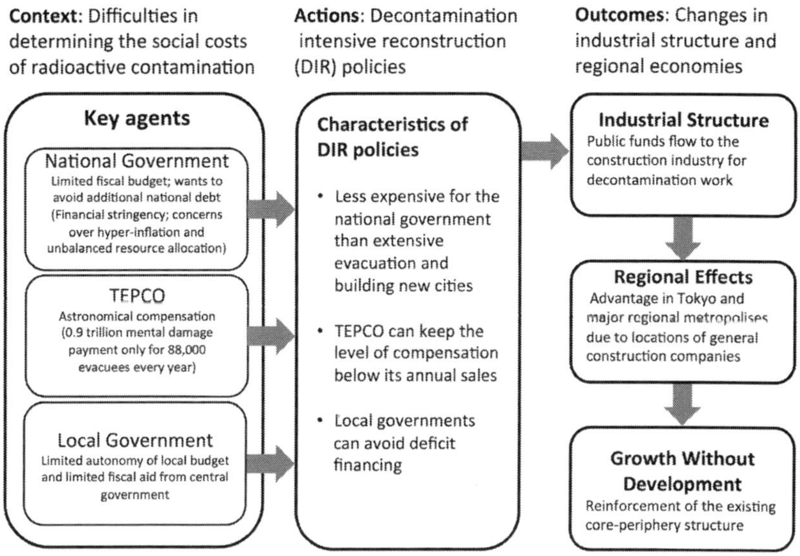

Figure 7.4 Structure of deconstruction–intensive reconstruction policy.

Source: Author, based on Fujimoto (2015b).

Table 7.2 Planned special account for the disaster (2013)

Reconstruction project	*Budgetary appropriation*	*Provisional budget*
Reconstruction Agency (1–6)	290	40
1. Support for the victims	18	1
2. Reconstruction of the cities and towns	166	33
3. Support of existing firms and creation of new employment	30	0.7
4. Decontamination	72	5
5. Adjustment cost	1	0.1
6. Administrative cost	0.4	0.05
7. Special grant-in-aid	60	–
8. Reserve fund for Fukushima	60	8
9. Reserve for national debt	6	0.6
10. Nationwide protection against disasters	12	–
11. Others	8	0.02
Total	438	50

Unit: billion yen

Source: Reconstruction Agency, 2013.

to be 6 trillion yen per year, that would be required if central Fukushima were included in the mandatory evacuation zones. If one adds losses from property values, production, lifetime earnings, and health care to the cost of "mental damage" compensation, the cost differentials would be even greater. In short, putting 726 billion yen per year into the decontamination project is a cheap alternative that is also a politically easy move because it helps to preserve the status quo in the existing regional political economy in which the construction industry plays a pivotal role.

Economic geography of the decontamination project

As the push for decontamination has become stronger, new or expanded branch offices of general construction companies (*zenekon*) have been set up in Fukushima City and in other middle-sized cities, like Koriyama and Iwaki Cities, in the prefecture. At first glance, this development may be seen as one of the few positive effects of the otherwise devastating disaster, as it may create some employment. But the picture is not that simple.

The decontamination project involves complex transactions among construction-related companies of different sizes, roles, and locations. The headquarters of general construction companies, which receive decontamination contracts from central government, are usually located in large metropolitan areas such as Tokyo or Osaka. When general construction companies from metropolitan areas get a contract order from the government, they usually calculate the costs of decontamination work at the rate of 25,000 yen per day (250 US dollars) for each decontamination worker, and it is well known that they almost always subcontract their work to local subcontractors (Kajita 2001). These small- and medium-sized, first- or second-tier subcontractors—many of which are located in large regional metropolises and industrial areas along the Pacific Manufacturing Belt—usually get 15,000–20,000 yen (about 150–200 US dollars) per worker per day, although most of them are dummy companies to siphon off profits. Finally, actual decontamination work is undertaken by third- or fourth-tier subcontractors that operate locally in Fukushima and hire "real" workers with tools and equipment at around 12,000 yen per day (about 120 US dollars).

Because of the spatial structure of these multi-layered subcontracting processes, a large portion of the economic benefits of decontamination, including profits siphoned off from government contracts and their multiplier effects, go to the largest metropolitan areas stretching from Tokyo to Fukuoka. Although we do not yet have detailed input–output analyses of decontamination, economic benefits in the Tohoku region, especially in Fukushima, from actual operation of decontamination (i.e. labor earnings and their multiplier effects) are probably very limited.

In short, it has become apparent that the current decontamination-intensive reconstruction policy is reinforcing the three-tier structure of spatial divisions of labor, comprising the Pacific Manufacturing Belt, where most headquarter functions concentrate, at the top of the hierarchy, followed by second-tier cities (for example, Sendai City) with higher branch offices, and by smaller, third-tier cities (for example, Fukushima City) with the lowest-level branch offices. This hierarchy is also reflected in the locations of government agencies. Local offices of the Ministry of Environment and the Transport and Reconstruction Agency are located in Sendai City and Fukushima City. Both cities, characterized by extreme dependency on subsidies and grants from the central government in Tokyo, are public-investment gateways from the central government to local municipalities (cities, towns, and villages) that require decontamination projects. In the Japanese economic geographic literature, this regional economic situation has been referred to as a branch-office economy, characteristic of the Japanese regional structure in the postwar period (Fujimoto 2014) and has been criticized for its tendency to reinforce the economic dependency of peripheral regions on core regions, resulting in "growth without development" (Ando 1986).

Conclusions

This chapter has examined the financial constraints of the central government (and TEPCO) as a possible reason for the current geographical extent of evacuation zones in Fukushima following the nuclear accident and has argued that the government adopted the decontamination-intensive reconstruction policy to avoid the potential bankruptcy of the state. The main victims of this policy decision include the people and communities in central Fukushima (Naka-Dori), where the radioactive contamination level is relatively high (enough to be designated as an evacuation zone in Chernobyl), but whose large population would cause a state fiscal crisis if compensation were mandated.

Decontamination projects, as an alternative to more extensive evacuation, revive and reinforce the political economy that has long characterized the peripheral areas of Japan, where publicly financed construction work plays a key role in the regional economy. In fact, decontamination work is arguably worse than regular construction work because it is just the removal or relocation of plants, soil, or dust and does not produce any infrastructure. Central and local governments contract out decontamination work to private construction companies in cities that are hierarchically configured, and the existing hierarchical location patterns of construction-industry offices and government offices remain unchanged.

Ultimately, one of the central problems with the policies dealing with the Fukushima disaster was the severe deficiency of information disclosure and transparency about the decision-making processes of many essential policies, such as evacuation guidelines. No satisfactory rationale for using 20 mSv/year has been provided, nor do we know who actually came up with the guidelines. The public is expected simply to have faith in government decisions rather than getting full information to make informed decisions of their own. This fundamental problem is still being inadequately addressed in post-3.11 Japan.

Note

1 For example, the amount of government bonds was 44.2 trillion yen, nearly equal to tax and other revenues (42.3 trillion yen). Tax revenue plus government bonds equalled the total expenditure; therefore, half of the expenditure was financed by the bonds. The total amount of bonds and borrowings was calculated at 830 trillion yen, as of June 30, 2013, according to Japan's Ministry of Finance (MOF). Arguably Japan's macro-economy did not go bankrupt in the 2000s due to its public-debt dynamics, in which the savings from households and firms are largely capitalized as national bonds by way of many commercial banks to keep the fiscal balance equal (Ministry of Internal Affairs and Communications 2013).

References

Aldrich, Daniel P. 2011. "Future Fission: Why Japan Won't Abandon Nuclear Power." *Global Asia*, 6(2): 62–7.

Ando, Seiichi. 1986. *Chihou no Keizaigaku – Hatten Naki Seicho wo Koete* [Regional Economy – Beyond Growth Without Development], Tokyo: Nihonkeizaishinbunsha. [In Japanese.]

Cabinet Office. 2014. *Annual Report on the Japanese Economy and Public Finance (2014)*, Tokyo: Cabinet Office.

—— 2015. *National Accounts of Japan (2013)*, Tokyo: Cabinet Office.

Citizen's Radioactivity Measuring Station. 2013. "Location Map of the Recorded Doses." Accessed March 20, 2016. http://en.crms-jpn.org/cat/mrdata.html.

Elliott, David. 2013. *Fukushima: Impacts and Implications*, Basingstoke: Palgrave Macmillan.

Fujimoto, Noritsugu. 2014. "The Change of Hierarchy of Urban System: Analysis of Office Location Restructuring in Japan (2000–2010)." *Innovation and Supply Chain Management*, 8(1): 8–14.

—— 2015a. "Decontamination-oriented Reconstruction Policy in Fukushima Post 3.11: Real Versus Administrative Boundaries." *Disaster Advances*, 8(1): 53–8.

—— 2015b. "Government's Budget Constraint on the Fukushima Nuclear Disaster: Substitution Ratio of Decontamination to Evacuation in Japan." *Disaster Advances*, 8(9): 26–33.

Guha-Sapir, Debarati, Philippe Hoyois, and Regina Below. 2014. *Annual Disaster Statistical Review 2013: The Numbers and Trends*, Brussels: Centre for Research on the Epidemiology of Disasters (CRED).

Guha-Sapir, Debarati, Indhira Santos, and Alexandre Borde. 2013. *The Economic Impacts of Natural Disasters*. New York: Oxford University Press.

Guha-Sapir, Debarati, Femke Vos, Regina Below, and Sylvain Ponserre. 2012. *Annual Disaster Statistical Review 2011: The Numbers and Trends*, Brussels: Centre for Research on the Epidemiology of Disasters (CRED).

Johnson, Chalmers. 1982. *MITI and the Japanese Miracle*, Stanford, CA: Stanford University Press.

Kajita, Shin. 2001. "Public Investment as a Social Policy in Remote Rural Areas in Japan." *Geographical Review of Japan, Series B*, 74(2): 147–58.

Kobayashi, Yoshiaki. 1975. "The Inflation and Economic Crisis - Post War Japan." *The Doshisha Business Review*, 26(4–6): 314–49. [In Japanese.]

Lych, G. M. and Pateeva Z. G. 1999. "Chernobyl Catastrophe: Socio-Economic Problems and Ways to Address Them." *Proceedings of International Conference, Ten Years after the Chernobyl Catastrophe Conference*: 185–8.

Ministry of Internal Affairs and Communications. 2013. *White Paper on Local Public Finance*, Tokyo: Ministry of Internal Affairs and Communications.

Munro, Alistair. 2013. "The Economics of Nuclear Decontamination: Assessing Policy Options for the Management of Land around Fukushima Dai-ichi." *Environmental Science and Policy*, 33: 63–75.

Murakami, Yasusuke. 1996. *An Anticlassical Political-Economic Analysis: A Vision for the Next Century*, Stanford, CA: Stanford University Press.

Nuclear Regulation Authority. 2006. *Size of Radius Population around Nuclear Plants in Japan (2005)*, Tokyo: Nuclear Regulation Authority.

Rabl, Ari and Veronika A. Rabl. 2013. "External Costs of Nuclear: Greater or Less than Alternatives?" *Energy Policy*, 57: 575–84.

Reconstruction Agency. 2013. *Current Status and Path Toward Reconstruction*, Tokyo: Reconstruction Agency.

TEPCO. 2013. *TEPCO Illustrated*, Tokyo: TEPCO.

Voloshin, V. 1999. "Geographic Aspects of the Socio-Economic Consequences of the Chernobyl Catastrophe in Ukraine." *Proceedings of International Conference, Ten Years after the Chernobyl Catastrophe Conference*: 189–97.

8 Living with contamination

Alternative perspectives and lessons from the Marshall Islands

Sasha Davis and Jessica Hayes-Conroy

Introduction

The restoration and redevelopment of Fukushima after the earthquake, tsunami, and nuclear accident of March 2011 will be a difficult and complicated task. This process, however, can be aided by geographers and other scholars who are familiar not only with the local circumstances in the Tohoku region, but who are also familiar with the lessons of geographic scholarship in similar contexts. One of the important insights of research in other contaminated places is that people's experiences of contamination and disaster can be extremely varied depending on a person's identity and social location. Women experience contamination differently from men. Farmers experience contamination differently from urban residents. Parents experience it differently from children. Government planners experience it differently from schoolteachers. While most people in Fukushima are aware of this fact, these differences are often not taken seriously enough when planning for the clean-up and redevelopment of contaminated places.

In this chapter we argue that these differences must be taken seriously because they will have a profound effect on the success of future development initiatives in Fukushima. To demonstrate this, we have arranged our discussion in three parts. First, using information from Fukushima, we show how differences in identity and social location produce different conceptualizations of contamination and different strategies for coping with it. Second, we discuss two important theoretical perspectives in geographic scholarship—political ecology and feminist geography—which have developed out of decades of research by hundreds of scholars around the world. Much of the research that has led to these theories has been conducted in places where, like Fukushima, there are great differences between the way different people have experienced environmental dangers and social marginalization. Third, we discuss the empirical example of the restoration and redevelopment of the northern Marshall Islands, including Bikini Atoll, to examine how communities facing problems with nuclear contamination have coped with radioactive environments over decades.

Our hope is that people working to restore the landscapes and communities affected by the Fukushima Daiichi accident can learn from the experiences of these other researchers and communities. We believe the research traditions of political ecology and feminist geography have important contributions for planners and policy makers charting the future course for the Tohoku region. Furthermore, the experiences of the Marshall Islanders can serve to show how communities will respond in the long term to a nuclear disaster as well as show what strategies and plans do not work. We do not give this analysis of geographic theory for its own sake, or give an account of the Marshall Islands as if it is an isolated case. Instead, our aim is to use these examples to inform the practical work of decontamination and reconstruction so that life can be improved for the people of Fukushima.

How have people been impacted differently by the Fukushima disaster? Ethnographic research in Tohoku

In the summer of 2013, we engaged in a month of intensive ethnographic fieldwork in Fukushima Prefecture. During this time, we conducted over 40 interviews with a diverse range of residents: young and old, rural and urban, men and women, parents and those without children. If we had one "take away" message from this work, it was simply that the Fukushima disaster has impacted people variably. In this section, we provide a brief glimpse into this variation. We draw from our interview and participant observation data in order to illustrate how identity and social location impact both the way that people think about radiation contamination and the way that people go about dealing with it. We do so by way of four short examples: depression and isolation among the elderly, intergenerational food-safety concerns, monitoring and community vigilance among mothers, and the impacts of rural to urban migration on farmers.

The most commonly asked questions about the Fukushima disaster all deal with radiation. What levels of radiation are to be considered safe? Is the food okay to eat? Which areas are most contaminated? There are many people who debate these questions regularly, both scientists around the globe and local residents around Fukushima Prefecture. Nevertheless, radiation is not the only health concern to be associated with the Fukushima disaster. For many elderly people, feelings of depression and isolation are stronger and more tangible realities than the unknowns of low-dose-radiation exposure. Indeed, one of the strongest examples that we witnessed of how the disaster has impacted residents differently comes through the different perceptions of contamination of younger and older residents of Fukushima. While many young adults and children have decided to relocate due to radiation concerns, older generations often worry less about the impacts of radiation on their own bodies. To be elderly in the context of radiation means to be less susceptible to radiation, due to the calculation that older people will likely die of another cause before any radiation-specific illness could take hold. A common perception therefore is that young adults and children need to move away, to protect themselves from radiation, while older generations can remain in contaminated areas if needed, to care for what remains of family homes or businesses. There is a clear focus here on the direct bodily threats of radiation exposure, while secondary impacts on mental and physical health remain less of an immediate concern. The result of this, of course, is that families are split apart—an unfortunate but ostensibly necessary condition to keep young people safe. Older residents, meanwhile, are left behind to cope in isolation with feelings of depression, loneliness, and grief.

Related to the question of radiation risk is a frequently articulated concern over what is safe to eat. To be sure, the way that food safety is perceived and practiced in Fukushima varies greatly across the population. Over the course of our research we encountered all sorts of eaters: those who avoided all Fukushima produce while accepting local water, those who ate certain local vegetables but not others, those who avoided only mushrooms and mountain-foraged products, and those who specifically sought out Fukushima produce as a matter of support and solidarity. Although many factors of identity and social difference have an impact on these choices, age again appeared as one of the biggest influences. Many of our interviewees reported changes in family dynamics around mealtimes and, particularly, struggles between parents and grandparents regarding what is safe to feed young children. Rural families and those with long-standing gardening traditions seemed particularly susceptible to such conflicts. And, of course, a family's economic status also impacts decisions over what is okay to eat (and who is okay to eat it). Since Fukushima produce is now cheap to buy in stores and even more economical if grown at home, choices to eat locally or non-locally are also economic decisions. As a result of all of these influences, some families' cooking habits and childcare arrangements have shifted significantly. Families may eat together less, or cook more than one meal at each sitting. Mothers may take more care in arranging appropriate or acceptable meals, sometimes in conflict with the desires of grandmothers. There are also some individuals who believe it is important to eat foods that can help to cleanse radiation and toxins from the body, including foods high in fiber, pectin, and iodine. In sum, both within and between households, food decisions clearly differ substantially and often in ways that conflict both with past practices and concurrent, generationally driven behaviors.

A third example of how the Fukushima disaster has been experienced differently across the social landscape comes from monitoring and community vigilance. As anyone familiar with the Fukushima context will know, the prefecture has many localized "hotspots"—places where radiation readings are much higher than local air-monitoring stations are able to measure (see also Chapter 6). These hotspots might be localized to a particular park or rooftop and can even vary dramatically within one location. Local residents are commonly aware of the places where hotspots are likely to be; for example, radiation often collects where sediment does, at the base of a tree or a playground slide, or around a rain gutter or drainage basin. And residents are also able to test for such hotspots by borrowing monitoring equipment from their municipal government office. We spoke with many residents about radiation hotspots and monitoring, and we found that it was often young mothers who have taken up this task. Mothers in a sense have become citizen scientists, drawing from their learned knowledge of hotspots, using the borrowed equipment to test areas at risk, and sharing their discoveries with each other on social-media sites. Because much of the unpaid labor of childcare in Japan falls on mothers, this means that the Fukushima disaster has led to an increased workload for mothers, who must now operate as citizen scientists and food-safety experts as part of their parenting duties.

Our last example of uneven experience comes from the impacts of rural to urban migration. Undoubtedly, displacement is hard on anyone who experiences it, and the processes of evacuation in the aftermath of the Fukushima disaster were complex and challenging for all involved. Anyone who loses their community also loses their sense of place and, along with it, a part of their identity. Many of those we spoke with had experienced forced evacuation and articulated sentiments along these lines. But our sense was that the displacement process was particularly difficult for farmers and rural dwellers whose sense of self and life were built upon the land. Some of the farmers we spoke with had multi-generational ties to the land. All those who could stay on their land decided to remain, despite obvious hardships, and were determined to try to decontaminate. The others, who were forced to leave, have experienced extreme hardship due to the major life changes that rural to urban migration brings. One former farmer who is now living in "temporary" evacuated housing has gained over 60 pounds (27 kg) in two years. He explained to us that this weight gain is due to his sedentary lifestyle in the urban, evacuated-housing complex, so far removed from his own lands. This farmer also articulated a simultaneous loss of community; when we asked him why he didn't move to another rural location outside of Fukushima, he explained that the temporary housing complex also hosts what is left of his rural neighborhood. He and his family have friends at the same complex, and if he left, those ties would be broken as well.

These four examples are not meant to be comprehensive nor mutually exclusive. People's experiences and perceptions of the Fukushima disaster vary substantially but also overlap. They are complex and contradictory and not easily predictable, although they do seem to contain patterns inherent to social identity (for example, age, socio-economic status, parenthood, and geographic origin). Our overarching point is simply that this variation matters to understanding the true impacts of the disaster. Food-safety concerns are not the same concerns for all residents. Radiation risk does not mean the same thing for all people. Displacement is not evenly experienced. Recognizing the particularities and nuances of this variation is, we believe, the first step in being able to respond to such disasters in meaningful and effective ways.

What can systematize and explain some of this variation? Political ecology and feminist approaches in geography

We have found that theoretical perspectives in the discipline of geography can illuminate what is occurring in post-disaster Fukushima as well as inform the process of restoration and redevelopment of the region. In particular, two perspectives are enlightening in this context: political ecology and feminist geography. Both of these perspectives can help explain how different people experience disaster differently, but they do not merely recognize that these differences exist. Instead, they aim to analyze how these different experiences are socially and geographically structured. In other words, they examine how people's different experiences are *systematically* structured by the way the larger society is organized. In the case of political ecology, the emphasis is on how people's experiences and life-paths are affected by the physical environment and the larger political policies and social structures in which people find themselves. In feminist geography, the focus is on the way that people's experiences and life-paths are affected by their gender, age, economic class, and other attributes of identity. These research traditions are so expansive that it is not possible to give a full description of their nuances in the short amount of space we have in this chapter. Our aim in this section is merely to introduce these research approaches so that people involved in the reconstruction and redevelopment of the Tohoku region can consider looking more deeply into them. We believe this kind of scholarship can be helpful for creating better policies to meet the needs of people recovering from the Fukushima disaster.

Political ecology is a field of study which exists beyond, but also is largely influenced by, the discipline of geography. This perspective aims to situate environmental and social problems such as disasters in their larger environmental, political, economic, and social contexts (Rocheleau 2008; Neumann 2009; Robbins 2012; Kimura and Katano 2014). Eschewing apolitical perspectives that see disasters as only physical phenomena, scholarship in political ecology emphatically rejects the notion that disasters are "natural." Instead, political ecologists posit that while physical phenomena such as earthquakes, tsunamis, or hurricanes may be a trigger, the fact that these natural events kill, injure, and displace large numbers of people is mainly the result of human decisions and activities (Blaikie and Brookfield 1987; Pelling 2001; Birkenholtz 2011). Because of this, studies in political ecology stress that the impacts of disaster are not equal for individuals across a society. Instead these studies emphasize that there is a broad web of social and physical relationships that structure society and situate some individuals as more "at risk" than others, with impacts dispersed unevenly across the social landscape. In other words, an individual's specific experience of a disaster is largely determined by the particulars of their socio-spatial location and always contextualized within a larger political and economic structure. While this point has long been recognized in the political-ecology literature, it has become more widely accepted following Hurricane Katrina in New Orleans and the 2010 Haiti earthquake, where there were such stark contrasts in the way people of different races and classes were impacted by the disasters.

Historically, political ecology has been instrumental in illustrating how individuals are constrained in their ability to respond to crises by both the characteristics of the physical landscape around them and by larger social forces such as international capitalist markets, national state policies, and discourses about "proper" social roles within different landscapes (Neumann 2009; Schroeder 1999). There is also, however, an increasing recognition that this is not a one-way process, since landscapes and larger social forces are also shaped by individual and group actions (Fairhead and Leach 1996; Harris 2006; Robbins 2012). Earlier political-ecology research focused on relatively simple, vertical "chains of explanation" to show how human agency, landscapes, and social structures impact each other—usually with the arrows of causation originating from social structures, to show how the state, capitalism, or imperial discourses shape landscapes and restrict the agency of inhabitants. More recently, however, Dianne Rocheleau (2008: 724) has noted that within political ecology, "The center of gravity is moving from linear or simple vertical hierarchies (chains of explanation) to complex assemblages, webs of relation and 'rooted networks' ... with hierarchies embedded and entangled in horizontal as well as vertical linkages." In other words, the "arrows of causation" between social structure, human actions, and landscape are much more complex, as the three realms mutually affect and constitute each other. For example, Leila Harris' (2006: 187) work on the changing waterscapes of Anatolia illustrates how social differences (including those induced by gender, ethnicity, landholdings, and livelihoods) not only condition differential outcomes for residents of Anatolia, but also "are themselves fundamentally renegotiated and recast in relation to waterscape change." In a different but corresponding way, actor-network theorist Sarah Whatmore (2002) has illustrated the mutually constituting relationship between (physical) matter and (social) meaning through her examination of the "hybrid" geographies of natures, cultures, and spaces. Political-ecology scholars have also recently expanded the field's reach into issues of human health and the body, an obviously important focus for research on radioactive contamination. By examining the ways that human bodies are materially constructed by elements in the environment—and the ways environments are constituted in part by social processes and human actions—these scholars have addressed how issues of human health are strongly connected to environmental-management policies (Mayer 1996; Kearns and Moon 2002; Dyck 2003; King 2010).

Stemming from a post-structuralist philosophical perspective, this sort of approach to political-ecology research does not view landscapes, social structures, or individuals as having essential natures but rather demonstrates that each is literally (both materially and discursively) constructed out of the others (Braun and Wainwright 2001). Individual bodies and behaviors are quite literally shaped from the materials of the surrounding environment and are disciplined/molded toward certain responses by broader social structures. Landscapes are the result of human activity and the larger social policies governing them, as well as hydrogeological and ecological forces. Social structures are created and remade by acts of human agency and are powered in part by the productive power of landscapes (Giddens 1984).

Following this approach, we believe it can be helpful to view the landscapes, individuals, and social structures in post-disaster Tohoku as aspects of an interacting system that all affect each other. We feel there are a few benefits to this approach. First, it allows us to understand and analyze experiences of living in the contaminated region as both constituted by, and constitutive of, socio-spatial difference. In other words, a person's gender, age, and occupation affect their experiences of disasters, but these experiences of disaster also recreate what it means to be female, or old, or a farmer in contemporary society. Second, this approach also gives us an opportunity to examine and analyze instances when this two-way process has led to direct social actions aimed at influencing government policy and changing (decontaminating) the physical landscape. Third, this approach also gives us an important theoretical framework for analyzing the underlying logics of government policy and comparing these to the needs of various people living in the region. Put another way, the way suffering is experienced by people with different amounts of status and power (such as men versus women, older people versus younger, rural farmers versus urban consumers) affects how the government responds and how it structures its decontamination and compensation policies. Often this occurs in ways that leave the most vulnerable people in the region in the worst positions.

These concerns are more effectively addressed by combining political-ecology approaches with those from feminist geography. Feminist geography is a subdiscipline that looks specifically at the ways in which social differences—such as gender—are affected by politics, economy, culture, and the way physical and social landscapes are produced in relation to these social differences. Also, more recent feminist approaches have emphasized the centrality of people's bodies in any examination of human landscapes. It has become increasingly recognized that the particular materiality of bodies, as well as embodied drives and emotions, can motivate decision-making in a variety of human-environment contexts (Elmhirst 2011; Hayes-Conroy and Hayes-Conroy 2013; Nightingale 2011; Sultana 2011). This is a critical issue to take into consideration when looking at how individuals navigate contaminated landscapes: that is, how individuals experience and respond to the risk of radiation exposure through food and the local environment. Decisions about what to eat, where to live, or what to feed to family members are not *only* the result of scientific assessments or statistics of radiation levels, and neither are they *only* constructed socially through social position, discourses, and anecdotal evidence. They are both of the above, and they are *also* produced through embodied visceral drives such as familiarity, existing states of health, and taste preferences, as well as by emotions such as love, fear, shame, pride, and feelings of acceptance and belonging. In other words, when examining experiences of living in a contaminated environment we have to understand individuals not just as responding to social structures and discourses, but also as emotional and material bodies that make and respond to their own embodied demands in ways that can both reinforce and shift social structures and discourses as well as their physical landscapes (Hayes-Conroy and Hayes-Conroy 2013). Studying such embodied drives can help us to better understand individual experiences and decision-making in the face of various perceived health risks, and it can also help us to better understand how these decisions "scale-up" and affect landscapes and social structures more broadly. For example, the combined effect of the individual decisions that are made in response to perceptions of food and environmental safety can literally remake physical landscapes by dictating what places are able to viably produce particular agricultural products or be inhabited/occupied by particular individuals. Furthermore, as researchers have noted in a variety of contaminated sites around the world, the emotional demands of bodies also serve as motivations for political organizing and actions that have changed, sometimes radically, the operation of state power and international markets in a variety of places (Davis 2012; Hayes-Conroy 2009).

This last point is particularly critical not only from an academic perspective but a policymaking one as well. Intensive place-based political-ecology studies can be useful to managers, planners, and policy makers by demonstrating that what is going on at fine-scales (localities and individual bodies) is not only conditioned by larger social forces but also reflects back and has repercussions for the construction of larger social processes, patterns, institutions, and landscapes. By demonstrating that these personal experiences and reactions have far-reaching effects, we can also show that what is going on in Fukushima is not an entirely unique situation. In many respects, it is a situation shared by people living in contaminated places from Ukraine and Belarus to the Marshall Islands, Puerto Rico, Nevada, Russia, and (unfortunately) to future sites of radioactive and chemical contamination.

The Marshall Islands' experience with radiation, restoration, and redevelopment

In this section, we use insights from these theoretical approaches and combine them with consideration of other sites of extreme contamination where the authors have done research, particularly Bikini Atoll in the Marshall Islands. The purpose of this section is to analyze what has happened here to inform discussions about what may lie ahead for people in Fukushima. In the Marshall Islands, radioactive contamination has affected the landscape and marine environments for decades. The nature of the contamination on Bikini Atoll is a little different than Fukushima in that the radiation was spread across the landscape in a series of nuclear-bomb tests over years. Also, the sandy soils of the atoll do not hold onto cesium particles to the extent that the more clay-rich soils in Fukushima can. For this reason, on Bikini, the cesium contamination cycles through biological organisms more than it is held in soil (Simon 1997). There are also obvious differences in climate and vegetation. Despite these differences, the experience of people being displaced due to contamination and their efforts to restore and redevelop the area have some important lessons for the future of Fukushima.

Bikini was used as a nuclear-weapons test site from 1946 to 1958. During that time, 23 nuclear devices were detonated on the land, in the air, and in the sea near Bikini, and another 44 nuclear devices were detonated on nearby Enewetak Atoll. Of the 23 nuclear tests on Bikini, test shot "Bravo" on March 1, 1954 was the most destructive. It was a hydrogen-bomb test 750 times as explosive as the bomb detonated on Hiroshima. The bomb vaporized some of the islands on the northern rim of the atoll and left a large crater in the atoll reef. The pulverized bits of coral landed as fallout over the Pacific and on nearby populated Rongelap Atoll. A radioactive wave washed over Bikini "killing off all animal life except one hardy variety of rats" (Trumbull 1982: 49). The damage from Bravo was so great that there are reports the top American official in the Marshalls, Maynard Neas, warned Marshallese leaders that "If anyone breathes a word of this, they'll be shot before sunrise" (Johnson 1980: 57). The effects of the test were so significant that the date March 1 is now a national holiday in the Marshall Islands (Nuclear Victims' Day). Many residents of nearby Rongelap Atoll received high doses of radiation (1.9 sieverts). The Bravo test also contaminated the Japanese fishing boat *Daigo Fukuryu Maru*. One person on board, Aikichi Kuboyama, died from radiation poisoning and the tuna the crew caught entered the market in Japan and created a panic over "Bikini tuna" (Weideman 1954). In the wake of Bravo and subsequent tests, the atolls of Bikini, Enewetak, and Utrik were left uninhabited for decades. Shockingly, the people of Rongelap were returned to their atoll by the US military in 1957, only three years after the Bravo test and before the US ceased aerial nuclear testing in the Marshalls, in 1958. Many Rongelapese claim they were intentionally left in the contaminated environment as part of a US experiment to analyze the long-term effects of radiation exposure on an isolated population. They asked to be removed in 1985 but the US refused. The environmental group Greenpeace then helped relocate the islanders. The Rongelapese have had many radiation-related illnesses, especially high rates of thyroid cancer, which began to be seen about nine years after the Bravo explosions.

By 1968, the US government began proclaiming the safety of Bikini. They then began preparing to move former inhabitants back onto the atoll. The US tried to make the atoll habitable by clearing debris and planting close to 100,000 new palm trees across the island in the modernist style of strangely perfect rows. When the Bikinians visited their old home in the late 1960s with the thought of repatriation, they were saddened to see the damage that the island had sustained. Particularly disheartening was the vaporization of three of the islands in the Bravo test. Despite the damage, in August 1968 US President Lyndon Johnson officially proclaimed that the atoll was safe and that the Bikinians could return home (*Time* 1968). One article published at the time referred to Bikini as a "renovated paradise" (MacDougall 1974). The Bikinian local council, however, voted not to return due to distrust of the reports that the atoll was safe. They also stated, however, that they would not prevent individuals from going on their own if they desired (Niedenthal 1997). Land is of paramount importance to the status of Marshallese, and some Bikinians with large landholdings on Bikini decided to return to establish their old claims and live on them. Approximately 100 people moved back to the atoll. An article in *Life* magazine reported on the Bikinians' homecoming and their initial hesitancy to believe the atoll was safe (Mydans 1968). Tommy McCraw, a scientist with the US government, accompanied the Bikinians to reassure them of the islands' safety. Mydans (1968) writes:

> Farther down the beach we came upon a single, stunted coconut palm bearing stunted fruit. Two Bikinians were already there and had harvested some of the nuts and cut them open. On our arrival they held them out and asked if they were safe to drink. "Sure," said McCraw, "they're good," and he made a motion for them to drink. But they hesitated. One of them held his coconut out and made the same motion to McCraw. Laughing, he took the nut and drank from it till it was empty. The Bikinians laughed too and drank with confidence.

Within five years, many Bikinians were found to have extremely elevated amounts of radioactive cesium and strontium in their bodies: ten times and four times the determined safe limits respectively (Simon 1997). Scientists discovered the radioactive contamination resulted largely from eating coconuts. In 1978, those living on Bikini were once again expelled from their atoll. A Bikinian said of the experience:

> we really didn't have any worries until those scientists started talking about the island being poisoned again… We were so heartbroken that we did not know what to do… We were sad, but we didn't want to make a problem for the Americans. If they say move, we move.
>
> (Pero Joel, quoted in Niedenthal 2002)

Since the removal of people in 1978 there has been no permanent habitation of Bikini Atoll; however, it is frequently debated and discussed among members of the community, who are now scattered across the Marshall Islands and the US. As for nearby Enewetak and Rongelap, some people have moved back in the past decade after remediation efforts. In 1995 (41 years after Bravo), a comprehensive study was done to evaluate the possibility of people returning to Bikini. The study found that:

> Permanent resettlement of Bikini Island under the present radiological conditions without remedial measures is not recommended in view of the radiation doses that could potentially be received by inhabitants with a diet of entirely locally produced foodstuffs. This conclusion was reached on the basis that a diet made up entirely of locally produced food-which would contain some amount of residual radionuclides that could lead the hypothetical resettling population to be exposed to radiation from residual radionuclides in the island, mainly from cesium-137, resulting in annual effective dose levels of about 15 mSv (if the dose due to natural background radiation were added, this would result in an annual effective dose of about 17.4 mSv). This level was judged to require intervention of some kind for radiation protection purposes.
>
> (International Atomic Energy Agency, quoted in Niedenthal 2002)[1]

While the Bikinians considered returning under these circumstances, a distrust of US proclamations and controversy over different standards for US versus international sites led the Bikinians to reject returning to the atoll (Davis 2005). To this day, Bikini Atoll does not have any permanent inhabitants.

Conclusions: lessons from the Marshall Islands, political ecology, and feminist geography

What are the lessons of political ecology, feminist geography, and the Marshallese experience with radioactivity for people living in Fukushima and planning for its future? There are several. We will first address the lessons that can be learned from the Marshall Islands and then weave them together with the benefits of considering the perspectives of political ecology and feminist geography. By examining the case of the Marshall Islands, Fukushima residents should understand that the consequences of environmental radiation will affect the region for many decades. We should emphasize that this is not just because the cesium-137 half-lives are just over 30 years, but also because of the concern and stigma that will be attached to the region even where contamination levels are minimal. The second lesson to be learned is that one of the most difficult things about planning the restoration and redevelopment of an area with radioactive contamination is that people's responses to a given level of radiation can vary so much.

One of the primary lessons of the Marshallese experience with radiation is that there will be uncertainty about safety for a very long time, *even when scientific studies have accurate measures of radiation* (Davis 2005). This is because radiation measurements are ambiguous in many different ways. First, as most people in Fukushima are quite aware, there can be tremendous variations in levels of radioactivity over very small distances. Even if representative readings are known in certain places, other nearby hotspot areas may be much higher. Second, even when a radiation reading is known—for instance, 0.35 mSv/hour—what does that actually mean to the people traversing that landscape? Unfortunately, there is little scientific consensus about the consequences of long-term exposure to low doses of radiation. Third, even in cases where there is more certainty—like the link between iodine-131 exposure and thyroid cancer—the effects are probabilistic. People may know a given exposure increases risk, but by how much? Furthermore, even if a person's risk of cancer is increased by 1 in 1000, what does that actually mean to the person? Are they willing to take that risk, or have their children take that risk? In our interviews in Fukushima, just as in the Marshall Islands, it was very easy to find two people who analyzed risk very differently and considered a given radiation level to be either perfectly safe or incredibly dangerous.

The other lesson from the Marshall Islands is that when it comes to formulating policies or planning for redevelopment the different opinions of people in the area matter and have very real political and social effects. It is unwise to think of public sentiments that contradict scientific evidence as merely "irrational" or in need of "education." While it is important for policy makers to have reputable scientific information, it is also important to consider that nobody experiences or perceives radiation the same way. Mothers with schoolchildren may expect something very different from a peach farmer when the future of the region is being considered (and when acceptable levels of radiation are being established).

This is where the approaches of political ecology and feminist geography can also be helpful. These approaches recognize that people experience and interact with landscapes in very different ways, and they construct acceptable levels of risk in different ways. In other words, we believe it is important to examine risk through a "constructivist" view (Cutter 1993; Jasanoff 1999; Davis 2005). Whereas a "realist" perspective assumes risk is "a tangible by-product of actually occurring natural and social processes… [that] can be mapped and measured by knowledgeable experts, and, within limits, controlled," constructivists instead emphasize that risks "do not directly reflect natural reality but are refracted in every society through lenses shaped by history, politics and culture" (Jasanoff 1999: 137). According to this constructivist perspective, scientific studies may be conducted to determine the potential hazards of a location, but the findings of these studies are often only one voice in the chorus of narratives. Anecdotal information, dramatic events, and past experiences all impact perceptions of the dangerousness of a place as much as, or more than, reported scientific findings (Davis 2005). Constructivists would contend that rather than believing in one "true" level of scientifically verifiable risk, people conceive of risk by selectively focusing on some sources of information and disregarding others. Several studies highlight this by demonstrating that perceptions of risk caused by nuclear or chemical contaminants in the environment are amplified by prevailing social discourses and also vary strongly by gender, cultural group, education, class, and age (Bassett *et al.* 1996; Davidson and Freudenburg 1996; Finucane *et al.* 2000; Gustafson 1998; Pidgeon *et al.* 2003; McBeth and Oakes 1996; Phillips 2002; Adeola 2000; Driedger *et al.* 2002; Rondeau and McIntyre 2010). This has definitely been the case in the Marshall Islands, where scientific measurements of radiation are not nearly as important for determining a person's response to radiation risk as much as stories from relatives about the birth of deformed children, or the past experiences of being told something is safe when it was later determined that it was not. In the case of Fukushima, the source of information is also important in regards to trust. As TEPCO and central-government pronouncements are viewed by many as less credible than other sources, it is important that citizens' concerns—even if they are not scientifically "rational"—are considered in the process of setting standards, prioritizing areas for decontamination, and planning for future development. While people advocating for alternative standards may protest or mobilize through civil-society organizations, it remains a large challenge—in the Marshall Islands, Fukushima, and elsewhere—for differing opinions on radiation safety to be heard, and for those opinions to impact management practices. As Aya Kimura (2013) has shown in her discussion of the processes for setting standards of radiation for food in Fukushima, one potential mechanism for incorporating citizen perspectives has been to have multiple forums (governmental, consumer, and producer) where people can debate and be active in the process of formulating policies. The result of these multiple forums, however, can be different organizations calling for different standards with little discussion between the organizations that can bring about consensus.

Beyond radiation concerns there are also other important lessons from the Marshall Islands. One is the way that future generations will think of the region after years of displacement. Most Bikinians, especially younger ones, have no living memory of life on Bikini Atoll and have very little interest in "returning" to a place they widely view as "boring" and having no job opportunities. To younger generations, there is interest in visiting their contaminated homelands for cultural reasons but very little urge to permanently resettle there. Any redevelopment of the contaminated regions of Fukushima will also have to contend not only with fears of radiation but with doubts on the part of potential returnees that a successful and exciting future can be had in the relatively rural region.

Another phenomenon that we have seen in the Marshall Islands (as well as other sites of serious chemical contamination in Guam, Puerto Rico, and Hawaii) is a development option to minimize the need for decontamination by turning large tracts of formerly productive and human-occupied land into sites of alternative-energy projects or wildlife refuges. Some energy projects, like wind turbines and solar-panel arrays, can be situated in contaminated areas where other land uses are prohibited by the contamination. Also, turning radioactive or chemically contaminated areas into wildlife refuges has been very common. Bikini Atoll is a conserved area that has long hosted tourist visits (Davis 2007), and most of the military bombing range in Vieques, Puerto Rico was turned into a US Fish and Wildlife Refuge. Parts of Guam and Hawaii have experienced this as well. Many of the contaminated nuclear-bomb production sites around the continental US have also been turned into wildlife areas (Kirsch 2007; Krupar 2013). While preserving wildlife is a laudable goal, political ecologists point out that there are problems with contaminated places being turned into wildlife areas. First, it often serves to "preserve the contamination," because the nature-preserve designation is believed to mean that no decontamination is necessary (Davis *et al.* 2007). Another problem is that treating these places as "natural" landscapes erases the historical claims former residents have over the lands and abolishes the communities that once existed in these spaces.

In conclusion, there are many challenges for those planning the future of Fukushima. Restoration and redevelopment of radioactive areas is costly, time-consuming, and politically complicated. People in the region have different experiences of contamination and will therefore desire different outcomes from the process. Also, what is considered safe today may not be considered safe in 20 years. Over the span of a long-term project attitudes change, new information is gathered, and new findings about risk are determined. Despite these complications, there are resources that scholars, planners, and residents can draw upon to tackle this difficult task. People in the Marshall Islands, Ukraine, Belarus, US, Russia, and other areas of Japan who have long suffered nuclear contamination from bombs, reactor accidents, and leaking bomb-production facilities have experiences that people in Fukushima can learn from. In summary, these places have shown that conceptualizations of contamination vary widely among people, that scientific information is not the only source people consider when thinking about radiation safety, that nothing undermines trust like declaring something safe and then having to change that designation, and that even if a place is considered safe, getting young people to move back is still a difficult task. In these kinds of situation, it is important for scholars, planners, and policy makers to heed the lessons of political ecology and feminist geography and be sure to *hear* the different voices of people in the region and encourage them, with all of their varied perspectives, to participate in planning the future of Fukushima.

Note

1 The levels of radiation discussed here, of course, are lower than some of the "gray zone" areas considered suitable for habitation in Fukushima.

References

Adeola, Francis O. 2000. "Cross-national Environmental Injustice and Human Rights Issues: A Review of Evidence in the Developing World." *American Behavioral Scientist*, 43(4): 686–702.

Bassett, Gilbert W., Hank C. Jenkins-Smith, and Carol Silva. 1996. "On-site Storage of High Level Nuclear Waste: Attitudes and Perceptions of Local Residents." *Risk Analysis*, 16(3): 309–19.

Birkenholtz, Trevor. 2011. "Network Political Ecology: Method and Theory in Climate Change Vulnerability and Adaptation Research." *Progress in Human Geography*, 36(3): 295–315.

Blaikie, Piers and Harold Brookfield, eds. 1987. *Land Degradation and Society*, New York: Routledge.

Braun, Bruce and Joel Wainwright. 2001. "Nature, Poststructuralism, and Politics." In *Social Nature: Theory, Practice, and Politics*, edited by Noel Castree and Bruce Braun, 41–63. Malden, MA: Blackwell Publishers.

Cutter, Susan L. 1993. *Living with Risk: The Geography of Technological Hazards*, New York: Routledge.

Davidson, Debra J. and Wiluam R. Freudenburg. 1996. "Gender and Environmental Risk Concerns: A Review and Analysis of Available Research." *Environment and Behavior*, 28(3): 302–39.

Davis, Jeffrey Sasha. 2005. "'Is It Really Safe? That's What We Want to Know': Science, Stories and Dangerous Places." *Professional Geographer*, 57(2): 213–21.

—— 2007. "Scales of Eden: Conservation and Pristine Devastation on Bikini Atoll." *Environment and Planning D: Society and Space*, 25(2): 213–35.

—— 2012. "Repeating Islands of Resistance: Redefining Security in Militarized Landscapes." *Human Geography*, 5(1): 1–18.

Davis, Jeffrey Sasha, Jessica S. Hayes-Conroy and Victoria M. Jones. 2007. "Military Pollution and Natural Purity: Seeing Nature and Knowing Contamination in Vieques, Puerto Rico." *GeoJournal*, 69(3): 165–79.

Driedger, S. Michelle, John Eyles, Susan D. Elliott, and Donald C. Cole. 2002. "Constructing Scientific Authorities: Issue Framing of Chlorinated Disinfection Byproducts in Public Health." *Risk Analysis*, 22(4): 789–802.

Dyck, Isabel. 2003. "Feminism and Health Geography: Twin Tracks or Divergent Agendas?" *Gender, Place and Culture*, 10(4): 361–68.

Elmhirst, Rebecca. 2011. "Introducing New Feminist Political Ecologies." *Geoforum*, 42(2): 129–32.

Fairhead, James and Melissa Leach. 1996. *Misreading the African Landscape: Society and Ecology in a Forest-Savanna Mosaic*, Cambridge: Cambridge University Press.

Finucane, Melissa L., Paul Slovic, Chris K. Mertz, James Flynn, and Theresa A. Satterfield. 2000. "Gender, Race, and Perceived Risk: The 'White Male' Effect." *Health, Risk & Society*, 2(2): 159–72.

Giddens, Anthony. 1984. *The Constitution of Society: Outline of the Theory of Structuration*, Berkeley, CA: University of California Press.

Gustafson, Per E. 1998. "Gender Differences in Risk Perception: Theoretical and Methodological Perspectives." *Risk Analysis*, 18(6): 805–11.

Harris, Leila M. 2006. "Irrigation, Gender, and Social Geographies of the Changing Waterscapes of Southeastern Anatolia." *Environment and Planning D*, 24(2): 187–213.

Hayes-Conroy, Jessica. 2009. "Get Control of Yourselves! The Body as ObamaNation." *Environment and Planning A*, 41(5): 1020–5.

Hayes-Conroy, Jessica and Allison Hayes-Conroy. 2013. "Veggies and Visceralities: A Political Ecology of Food and Bodies." *Emotion, Space and Society*, 6: 81–90.

Jasanoff, Sheila. 1999. "The Songlines of Risk." *Environmental Values*, 8(2): 135–52.

Johnson, Giff. 1980. "Nuclear Legacy: Islands Laid Waste." *Oceans*, 13: 57–60.

Kearns, Robin and Graham Moon. 2002. "From Medical to Health Geography: Novelty, Place and Theory After a Decade of Change." *Progress in Human Geography*, 26(5): 605–25.

Kimura, Aya Hirata. 2013. "Standards as Hybrid Forum: Comparison of the Post-Fukushima Radiation Standards by a Consumer Cooperative, the Private Sector, and the Japanese Government." *International Journal of Sociology of Agriculture & Food*, 20(1): 11–29.

Kimura, Aya Hirata and Yohei Katano. 2014. "Farming After the Fukushima Accident: A Feminist Political Ecology Analysis of Organic Agriculture." *Journal of Rural Studies* 34: 108–116.

King, Brian. 2010. "Political Ecologies of Health." *Progress in Human Geography*, 34(1): 38–55.

Kirsch, Scott. 2007. "Ecologists and the Experimental Landscape: The Nature of Science at the US Department of Energy's Savannah River Site." *Cultural Geographies*, 14(4): 485–510.

Krupar, Shiloh R. 2013. *Hot Spotter's Report: Military Fables of Toxic Waste*, Minneapolis, MN: University of Minnesota Press.

McBeth, Mark K. and Ann S. Oakes. 1996. "Citizen Perceptions of Risks Associated with Moving Radiological Waste." *Risk Analysis*, 16(3): 421–7.

MacDougall, W. 1974. "Twenty-eight Years after Atom Blast, the Bikinians Return Home." *U.S. News and World Report*, 77, December 16: 52–4.

Mayer, Jonathan D. 1996. "The Political Ecology of Disease as One New Focus for Medical Geography." *Progress in Human Geography*, 20(4): 441–56.

Mydans, Carl. 1968. "Twenty-two Years—Twenty-three Blasts Later: Return to Bikini." *Life*, October 18.

Neumann, Roderick P. 2009. "Political Ecology: Theorizing Scale." *Progress in Human Geography*, 33(3): 398–406.

Niedenthal, Jack. 1997. "A History of the People of Bikini following Nuclear Weapons Testing in the Marshall Islands: With Recollections and Views of Elders of Bikini Atoll." *Health Physics*, 73(1): 28–36.

Niedenthal, Jack. 2002. *For the Good of Mankind: A History of the People of Bikini and Their Islands*. 2nd ed., Majuro Marshall Islands: Bravo Publishing.

Nightingale, Andrea J. 2011. "Bounding Difference: Intersectionality and the Material Production of Gender, Caste, Class and Environment in Nepal." *Geoforum*, 42(2): 153–62.

Pelling, Mark. 2001. "Natural Disasters?" In *Social Nature: Theory, Practice, and Politics*, edited by Noel Castree and Bruce Braun, 170–88. Malden, MA: Blackwell Publishers.

Phillips, Sarah D. 2002. "Half-lives and Healthy Bodies: Discourses on 'Contaminated' Food and Healing in Post-Chernobyl Ukraine." *Food and Foodways*, 10(1–2): 27–53.

Pidgeon, Nick, Roger E. Kasperson and Paul Slovic. 2003. *The Social Amplification of Risk*, Cambridge: Cambridge University Press.

Robbins, Paul. 2012. *Political Ecology: A Critical Introduction*, Malden, MA: Blackwell Publishers.

Rocheleau, Dianne E. 2008. "Political Ecology in the Key of Policy: From Chains of Explanation to Webs of Relation." *Geoforum*, 39(2): 716–27.

Rondeau, Krista and Lynn McIntyre. 2010. "'I Know What's Gone Into It': Canadian Farmwomen's Conceptualisation of Food Safety." *Health, Risk & Society*, 12(3): 211–29.

Schroeder, Richard A. 1999. *Shady Practices: Agroforestry and Gender Politics in the Gambia*, Berkeley, CA: University of California Press.

Simon, Steven L. 1997. "A Brief History of People and Events Related to Atomic Testing in the Marshall Islands." *Health Physics*, 73(1): 5–20.

Sultana, Farhana. 2011. "Suffering for Water, Suffering from Water: Emotional Geographies of Resource Access, Control, and Conflict." *Geoforum*, 42(2): 129–32.

Time. 1968. Home to Bikini. August 23: 17.

Trumball, R. 1982. "An Island People Still Exiled by Nuclear Age." *U.S. News and World Report*, 93, October 18: 48–50.

Weideman, Elizabeth. 1954. "Ashes of Death: First H-bomb Victims." *Nation*, October 9.

Whatmore, Sarah. 2002. *Hybrid Geographies: Natures Cultures Spaces*, Thousand Oaks, CA: Sage.

9 Radioactive contamination of forest commons

Impairment of minor subsistence practices as an overlooked obstacle to recovery in the evacuated areas

Hiroyuki Kaneko

This chapter identifies and addresses a problematic limitation of current measures for bringing residents back to areas evacuated as a result of the Fukushima Daiichi nuclear disaster: namely, the exclusion of forested lands from decontamination and compensation measures. While a narrow economic calculus renders communal forest spaces, and the "minor subsistence" practices pursued within these spaces, "invisible" and thus excluded from key recovery policies, I demonstrate that forests are of great importance from the perspective of village life and sociality and thus vital to recovery in formerly evacuated areas.

Returning evacuees and the limits of decontamination and compensation

The nuclear accident at Fukushima Daiichi resulted in the compulsory evacuation of 12 municipalities of Futaba and Soma Districts (Figure 9.1). As the nuclear disaster unfolded in the days following March 11, 2011, residents of the communities near the nuclear plant were forced to collect their belongings and abruptly flee in a highly disordered evacuation. Today, a large number of these residents remain in evacuation. Over the course of the last few years, the evacuated areas have been reorganized into a matrix of different evacuation zones (Figure 9.1). In the zones nearest to the plant, entrance remains restricted and these areas will continue to remain off-limits well into the future. In other areas, however, evacuation orders have been lifted or are expected to be lifted, and various efforts are being made to bring residents back to their former communities. In some communities residents have already begun to return, but not without hardships and controversy.

While recovery policies are oriented toward residents returning to the evacuated areas, these policies have drawn criticism from both evacuees and academics. For example, Masamura (2013) argues that recovery policies oriented toward quickly returning residents to the evacuated areas too heavily stress the notion of "permanent residence" and thus tend to exclude other options, such as relocation of communities to other sites or a slower, less rushed return process. From this

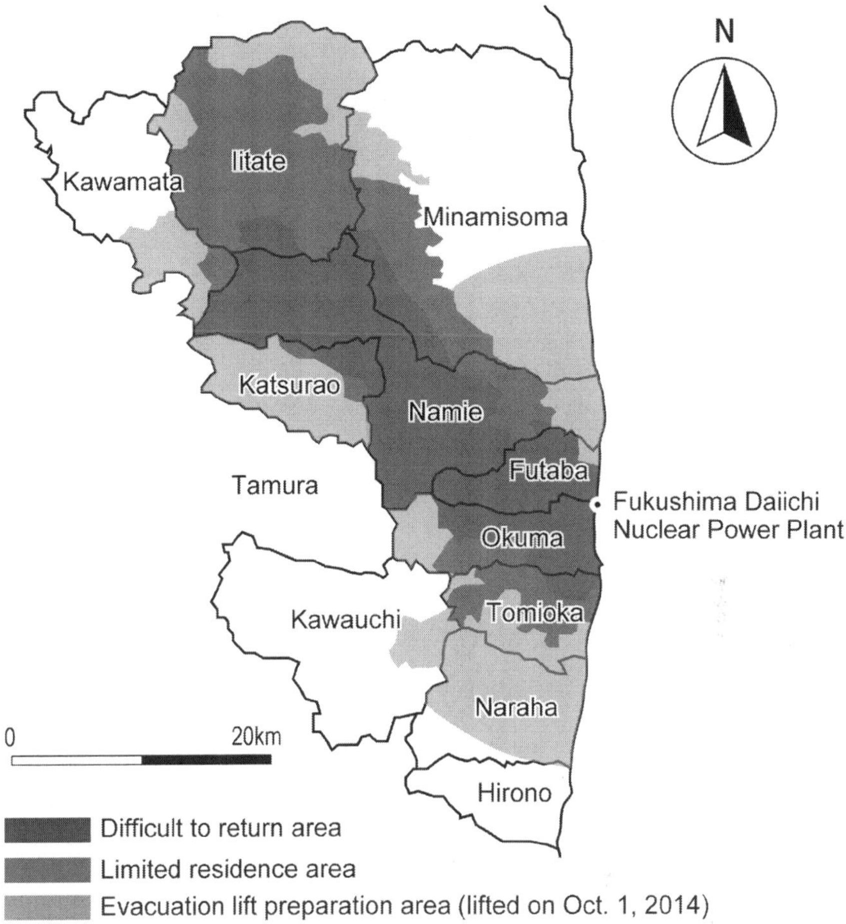

Figure 9.1 Evacuation designations in the area around Fukushima Daiichi as of September 2014.

Source: Author, based on Ministry of Economy, Trade, and Industry (2014).

perspective, recovery plans assuming the return of residents to the evacuated areas deny evacuees their "right to evacuate." This is an extremely important insight, and one that must not be overlooked or under-emphasized. Yet, at the same time, if government recovery policies were oriented solely toward relocation of evacuated communities to new sites, then the many residents hoping to return to the evacuated areas would be denied their "right to return." Accordingly, as argued by Seki (2013), it is imperative that recovery policies ensure both evacuees' right to return to their former communities and their right to remain in evacuation.

While conscious of the need to ensure the right to evacuate, this chapter focuses on how to improve policies for promoting the timely return of residents to the evacuated areas for one basic but crucial reason: many evacuees wish to return home.[1] The results of a questionnaire survey conducted by Seki (2012) in Nahara Town, one of the communities evacuated in the aftermath of the nuclear accident, indicate that approximately 40 percent of residents wished to return to their evacuated community. These results may seem somewhat surprising at first. It seems rather improbable that 40 percent of the residents of a community located 10–20 km from the epicenter of an unresolved nuclear disaster would wish to return home. For many readers of this chapter—academics accustomed to frequently moving to pursue education and employment opportunities—this strong desire to return to the afflicted communities can be somewhat difficult to comprehend. Faced with a similar disaster near their own community, many readers might be determined to secure financial compensation and begin their lives anew in an area free from hardships and health risks. However, in rural agricultural communities like those around the Fukushima Daiichi Nuclear Power Plant, residents have deep ties to landscape and community, and their ways of life and livelihood are firmly bound in place. Accordingly, it is imperative to ensure that these residents are able to return home through the implementation of appropriate measures.

Without question, the key measure for bringing residents back to the evacuated areas is decontamination. In 2011, the national government passed the Act on Special Measures concerning the Handling of Pollution by Radioactive Materials to establish a legal framework for advancing decontamination measures. This act stipulates that evacuation designations can only be lifted after the air radiation dose rate of an area has been confirmed to be under 1 mSv/year. The actual procedures to be performed are outlined in the Decontamination Guidelines published by the Ministry of the Environment (2013), the agency placed in charge of decontamination operations. In total, 99 municipalities in eastern Japan are targeted for decontamination, with operations focusing predominantly on Futaba and Soma Districts adjacent to Fukushima Daiichi (Figure 6.2, in Chapter 6). By April 2015, decontamination operations had been initiated in over 80 percent of the localities where the return of residents is expected in the future (i.e. in all of the evacuated areas outside the difficult-to-return zones).

The Ministry of the Environment's Decontamination Guidelines provide a critical framework for advancing decontamination. However, they also reveal how a detached "view from above" enables the exclusion of certain spaces—namely, forests—from decontamination targets and efforts. To ensure basic health and safety, the Decontamination Guidelines prioritize the "area of life activity" of contaminated communities. This "area of life activity" is essentially comprised of two zones: residential and agricultural environments. The problem here is that, as described by Fukuta (1989), the "life activity" of agricultural villages is actually composed of a tripartite spatio-ecological structure consisting of *mura* (i.e. residential village spaces), *nora* (i.e. agricultural fields) and *yama* (i.e. common forests and grasslands) (Figure 9.2). The importance here of this tripartite model of agricultural life activity is that it brings to light the spatial limits of decontamination. While the residential *mura* and agricultural *nora* are targeted for decontamination, the forest commons of the *yama* remain excluded from these efforts.

This is for two main reasons. First, the forested *yama* landscape is more spatially expansive than the residential *mura* or agricultural *nora*, and decontamination of this expansive area presents serious technical and financial difficulties. A second reason flows from the conceptualization of "life activity" advanced in the Decontamination Guidelines. When the life and livelihoods of agricultural communities are evaluated from an objective perspective focused on basic survival and efficient agricultural production, the *yama* landscape is marginalized as superfluous. This is due to the fact that, at present, forests are not a primary component of the commercial economy of many rural agricultural communities. Although historically important, the decline of forestry has rendered the *yama* landscape dispensable when evaluated with a purely economic calculus focused on agricultural output. In short, village forests have been excluded from decontamination to secure cost efficiency and political expediency through a strictly commercial conceptualization of agricultural "life activity" that overlooks the importance of forests to village life and sociality.

For many of the same reasons, village forests have also been excluded from another key component of recovery: compensation. While compensation for damages resulting from the nuclear disaster may not be a form of recovery policy in a strict sense, it is a crucial component of the overall recovery package necessary for enabling disaster-afflicted communities to overcome their deep and continuing losses. In total, nine types of damage stemming from the nuclear disaster are eligible for compensation.[2] Forests are, in general, eligible for compensation under the broad category of damages to agricultural, forestry, and fisheries products. However, while eligible for compensation, the fact is that commercial forestry in the area, as in much of Japan, is faced with serious economic challenges, and compensation for damages to commercial forestry in the area is limited. Moreover, while commercial forest products are eligible for compensation, these products are only one portion of the stream of goods and services that residents gather

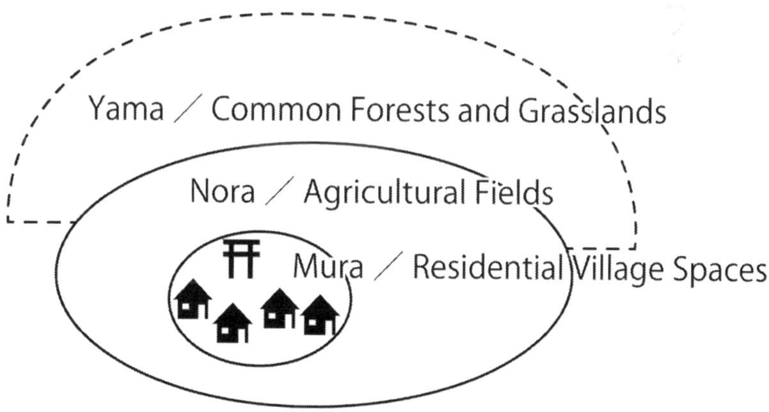

Figure 9.2 Tripartite structure of agricultural villages.

Source: Author, based on Fukuta (1989) and Ikegami (2007).

from the forest. Non-timber forest products procured through hunting, fishing, and gathering have been heavily contaminated as result of the nuclear disaster but are not eligible for compensation because they are primarily for household and community subsistence. To obtain compensation, all damages must be verified through the provision of objective proof, such as contracts or prior receipts. The issue here is that, as with decontamination, a commercial logic excludes the importance of forests to village life and sociality. In short, the commercial logic of both compensation and decontamination renders the spaces and practices of village forests invisible.

The problem is that the above conceptualization of village forests as dispensable is fundamentally flawed from the perspective of village life and sociality. While the commercial importance of the *yama* landscape has indeed declined, the importance of village forests has never been limited to commercial interests. Villagers have long pursued numerous non-commercial forestry management and subsistence practices. Following Matsui (1998, 2004), I would like to conceptualize these activities as forms of "minor subsistence," a term that speaks to the many and various forms of subsistence that are often overshadowed by emphasis on the major subsistence activity of a particular society. On the one hand, the concept of minor subsistence speaks to the importance of hunting, fishing, and gathering as a calorific supplement for agricultural livelihoods. On the other, Matsui coined this term to highlight the importance of minor subsistence practices as a critical source of individual and communal identity and wellbeing. In what follows, I will highlight the social importance of minor subsistence practices by identifying how products procured from village forests circulate through communities, socio-materially binding the community together through a chain of practices extending from management to gathering, sharing, and final consumption. Excluding village forests impairs recovery and inflicts continuing damage on the nuclear-disaster-afflicted communities of Fukushima. In order to improve recovery policies for the nuclear-disaster areas it is necessary to understand why residents see the contamination of the *yama* landscape and the continuing exclusion of this landscape from decontamination policy and practice as harmful to village life and recovery.

The remainder of this chapter presents a case study of Kawauchi Village, the first comprehensively evacuated municipality to issue a call for residents to return and, as such, an important test case for the feasibility of return more broadly. Fieldwork and archival research conducted in Kawauchi included analysis of the effects of the disaster on the community, historical research, and interviews on the historical utilization and importance of forests, as well as interviews to identify the many issues associated with utilizing the forest commons of the village after the nuclear disaster. The following section provides an overview of Kawauchi Village, highlighting the impacts of the disaster on the community and the progress of returning evacuees. I will then analyze the pre-2011 history of the *yama* landscape both in terms of a visible history of forestry and a less visible history of minor subsistence practices. The subsequent section provides a case study of the sociality of mushrooming in the village that shows that the forests of the village are a commons and the resources gathered from the forests circulate through the community and socio-materially tie it together. Lastly, I look at minor subsistence practices after the earthquake and nuclear disasters. It demonstrates that minor subsistence practices have been one of the major draws for villagers to return but also a continuing source of incomplete recovery, antagonism, and affliction. The conclusion advocates for further academic attention and concrete policy for the overlooked *yama* landscape.

Kawauchi Village: a canary in a coal mine for returning evacuees

Kawauchi Village is an isolated rural community. However, it has become an important test case—a canary in a coal mine—for evaluating the feasibility of returning residents to the evacuated areas of Fukushima. The following section provides an overview of the village, focusing on the impacts of the earthquake and nuclear disaster on the community and the current progress of returning residents to this former evacuated area.

General overview of Kawauchi and the impacts of the nuclear disaster

Kawauchi is an isolated rural community that has much in common with other "mountainous rural districts" throughout Japan. Of the total village land area of 19,738 ha, approximately 86 percent (17,023 ha) is forests. The fact that agricultural land totals only 970 ha, or 5 percent of the village land area, is indicative of the expansiveness of forest cover in the village. The pre-earthquake population of Kawauchi was approximately 3,000. Most residents are involved, at some level, with agriculture, including rice and other products. Like other mountainous rural districts throughout Japan, the village had experienced both aging and population loss well before the earthquake.

Generally speaking, the impacts of the initial earthquake disaster of March 11, 2011 on the village were relatively minor. Although shaking from the earthquake in the village measured a 6-upper on the Japan Meteorological Agency seismic-intensity scale (Chapter 1), no lives were lost as a result of the initial earthquake. Damage to buildings included eight totally destroyed structures, 512 partially destroyed structures and 160 partially damaged structures. Kawauchi was impacted during the immediate disaster by the earthquake only, and thus damages to the community were relatively minor, at least when compared to the devastation inflicted by the tsunami on communities along the coast.

However, the series of hydrogen explosions at Fukushima Daiichi brought Kawauchi into abrupt crisis. Approximately one third of the village land area is located with 20 km of Fukushima Daiichi, and the remainder of the village is located within 30 km of the plant (Figure 9.3). Accordingly, a comprehensive and mandatory evacuation of the entire village was initiated on March 15, 2011. The village offices were moved 40 km away to Koriyama City and most residents were evacuated to Koriyama.

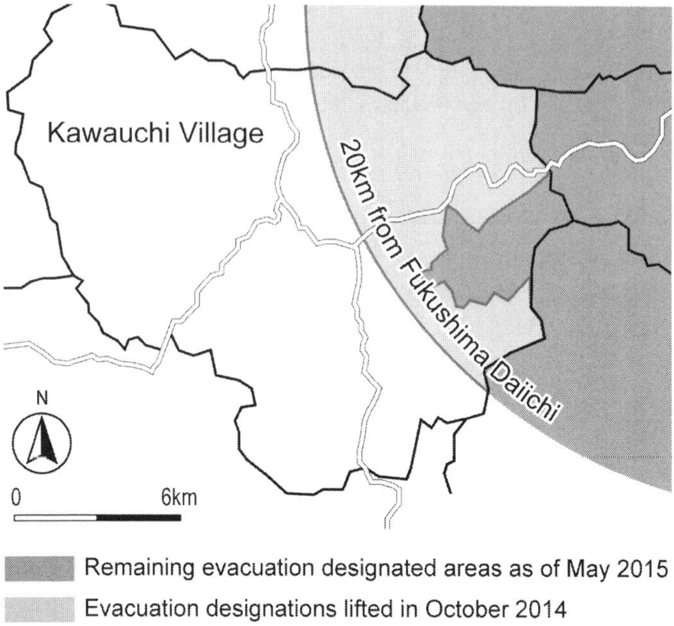

Figure 9.3 Evacuation designations in Kawauchi Village.

Source: Author.

The call to return and the progress of return

As mentioned, the attention of other evacuated communities, experts, and the general public was drawn to the isolated rural community of Kawauchi by the call on January 31, 2012 for residents to return to the village. This call to return has made Kawauchi an important test case for residents of other evacuated areas around Fukushima Daiichi who hope to return to their evacuated homes and communities and for policy makers intent on understanding the many issues involved with returning residents to evacuated areas.

Decontamination measures in Kawauchi were initiated on November 11, 2011 and completed on March 31, 2014. In accordance with the Decontamination Guidelines issued by the Ministry of the Environment, these efforts focused on the residential and agricultural areas of the village. The forested areas of the village were not targeted for decontamination. Even without conducting costly, difficult, and seemingly unnecessary decontamination of the forested *yama* landscape of the village, the air radiation dose rate fell below the objective standard set by the government, and villagers were able to return.

Residents began to return to Kawauchi in April 2012. As of January 2015, 639 residents, or approximately one fifth of the pre-earthquake population of the village, had returned to Kawauchi. At present, 1,158 residents remain in evacuation in Koriyama City but actively commute back and forth from Kawauchi. Thus, in total, nearly 2,000 residents have re-established ties to their former community.

Major forestry and minor subsistence: the *yama* landscape before 2011

Drawing on archival research and in-depth interviews conducted in Kawauchi, this section provides a broad overview of the history of forest utilization in the village prior to the nuclear disaster. For centuries the forests of the village were the site of both a major forest industry and various minor subsistence practices. As I will show, forestry in the village went into abrupt decline with the arrival of the "fuel revolution" of the 1960s. However, even as the forestry industry declined and forests were marginalized from the commercial ledger of Kawauchi, the forest commons remained socially important as sites for minor subsistence.

The charcoal boom years

Kawauchi Village was well known throughout Japan as a forestry region until the 1960s. As noted, forests occupy approximately 86 percent of the village. In local history books the prosperity derived from these rich forests is expressed via the phrase "charcoal kingdom," a term used to describe the boom years of charcoal production in the community. During these years, Kawauchi was the top producer of charcoal in all Japan, and charcoal was the most important product of the village.

In the 1960, however, the prosperity of the "charcoal kingdom" came to an abrupt end for two main reasons. First, the forests of the village had been over-utilized. The second, and more important reason, was the broader "fuel revolution" that swept Japan and transformed energy utilization in the country during the 1960s. As the availability of power sources such as electricity and gas spread throughout the nation, demand for charcoal abruptly evaporated, resulting in the devaluation of former charcoal-producing forests throughout the country and in Kawauchi.

As wood energy was falling into decline, another form of energy appeared on the scene in Fukushima: nuclear energy. The energy revolution was delayed in Fukushima and, in the 1960s, efforts were initiated to attract nuclear energy plants to the region. This was at a time when nuclear energy was attracting much attention throughout Japan as a new form of power generation. A power-generating company found a disused airfield in Okuma and Futaba. In 1961, a joint cooperation proposal was accepted by Okuma and Futaba councils, and plans were set for the construction of Fukushima Daiichi. Land was purchased, the plant constructed, and, on March 26, 1971, reactor unit 1 began operating. Kawauchi drew benefits from this energy transition and the age of wood energy was firmly in the past.

Sub-histories of the forest: seasonal minor subsistence

Activities conducted in the *yama* landscape were not, however, limited to commercial forestry. Various hunting and gathering activities to support self-subsistence were conducted in the *yama* landscape of the village. The minor subsistence practices of Kawauchi shift with the seasons. In spring, a number of *sansai* (wild edibles) are gathered (Table 9.1). Particularly favored among these wildly harvested edibles are bracken shoots. Most of these wild edibles are prepared through a process of pickling before consumption. Next in the seasonal procession of hunting and gathering is fishing in local streams during the summer months. The most prized fish in Kawauchi is char. Fishing in the area was in decline well before the earthquake disaster. As a result of agricultural pesticide usage, pollution from sewers, and environmental transformations caused by river-improvement projects, fishing in the village was radically transformed. However, char populations were relatively unaffected by these changes. Indeed, char was the only fish that could still be caught after river-improvement projects were completed. When the fishing season is complete, and autumn has arrived, the mushroom-gathering season begins (Figure 9.4). More than 20 mushroom species are gathered in Kawauchi. The nuts of trees are also gathered in autumn. Hunting is conducted in winter. In the days when the charcoal business was booming in Kawauchi, traps were laid for

Table 9.1 The products of the *yama* landscape

	Japanese (folk-term)	*English*	*Scientific (Latin)*
Spring (wild edible plants)	fuki	butterbur	*Petasites japonicus*
	fuki-no-to	butterbur flower shoots	*Petasites japonicus*
	warabi	bracken shoots	*Pteridium aquilinum*
	zenmai	Japanese royal fern	*Osmunda japonica*
	ukogi	fiveleaf aralia	*Eleutherococcus sp.*
	kawazenmai	river Japanese royal fern	*aquatic Osmunda japonica*
	kuzu	kudzu	*Pueraria thunbergiana*
	taranome	Japanese angelica-tree shoots	*Aralia elata*
	koshi-abura	young leaves of koshi-abura	*Eleutherococcus sciadophylloides*
	momijigasa	young leaves and shoots of momijigasa	*Cacolia delphiniifolia*
Summer (fishing)	kachika	Japanese fluvial sculpin	*Cottus pollux*
	ugui	big-scaled redfin	*Tribolodon hakonensis*
	dojyo	loach	
	funa	crucian carp	*Carassius sp.*
	iwana	char	*Salvelinus sp.*
		masu salmon	*Oncorhynchus masou*
	ayu	sweetfish	*Plecoglossus altivelis*

(Continued)

Table 9.1 (Continued)

	Japanese (folk-term)	*English*	*Scientific (Latin)*
Autumn (nuts and mushrooms)	tochinomi	Japanese horse-chestnut	*Aesculus turbinata*
	kuri	chestnut	*Castanea crenata*
	kurumi	walnuts	*Juglans*
	akebi	chocolate Vin	*Akebia quinata*
	mamedago	truffle	*Rhizopogon roseolus*
	amitake	suillus bovinus	*Suillus bovinus*
	chitake	tawny milkcap mushroom	*Lactarius volemus (Fr.: Fr.) Fr.*
	rokusyo	lactarius hatsudake	*Lactarius hatsudake Nobuj. Tanaka*
	honshimeji	species of edible mushroom	*Lyophyllum shimeji*
	kakishimeji		*Cortinarius tenuipes (Hongo) Hongo*
	sakurashimeji		*Hygrophorus russula (Schaeff. : Fr.) Quél*
	kusashimeji		*Lyophyllum decastes*
	murasakishimeji	blewit	*Lepista nuda*
	matsutake	matsutake mushroom	*Tricholoma matsutake (S.Ito et Imai) Sing.*
	maitake	hen of the woods	*Grifola frondosa*
	naranokimodashi	armillaria mellea	*Armillaria mellea subsp. Nipponica*
	kurinokimodashi		*Hypholoma sublateritium (Fr.) Quél*
	inohana		*Sarcodon aspratus (Berk.) S. Ito*
	koganetake	golden cap	*Phaeolepiota aurea (Matt.:Fr.) Maire.*
	houkidake		*Ramaria botrytis (Pers.: Fr.) Ricken*
	mukitake	olive oyster	*Sarcomyxa serotina*
	shiitake	shiitake mushroom	*Lentinula edodes (Berk.) Pegler*
	usutake	woolly chanterelle	*Gomph us floccosus Schw. (Singer)*
	ushikodake		*Boletopsis leucomelas*
	yukinoshita	flammulina velutipes	*Flammulina velutipes (Curt.: Fr.) Sing.*
Winter (hunting)	yamadori	copper pheasant	*Syrmaticus soemmerringii*
	tsugumi	dusky thrush	*Turdus eunomus*
	kiji	green pheasant	*Phasianus versicolo*
	inoshishi	wild boar	*Sus scrofa leucomystax*
	shika	deer	*Cervus nippon*
	yamausagi	wild rabbit	*Lepus brachyurus*

Source: Author, based on fieldwork.

Figure 9.4 Mushrooms gathered from the forests of Kawauchi (pre-earthquake: c. 2000).

copper pheasant and wild rabbit, and these traps were then checked and game collected when villagers returned to the village from the forest. At the peak there were 190 gun hunters in the village. Now, after the earthquake, there are fewer than ten.

As evidenced by the above analysis, the *yama* landscape of Kawauchi Village has been utilized for minor subsistence activities. These resources have been utilized as foods for home consumption and have greatly enriched the diet of the community. From this perspective it can be said that minor subsistence has been a vital and indispensable component of life in Kawauchi. However, minor subsistence practices are also socially vital.

The gathering and circulation of mushrooms: the *yama* as social commons

We turn our attention here to a case study of mushrooms and mushrooming in Kawauchi in order to highlight the broader social import of minor subsistence activities conducted in the *yama* landscape in the village, as well as the common consciousness of this landscape that sustains, regulates, and circulates through the community. Minor subsistence has not only been important as a calorific supplement to village diets but also as a space that sustains social relations and communal consciousness. The *yama* landscape is not a commons from a land-ownership perspective. It has been divided through a complicated historical process into national forests owned and managed by the national government, village forests owned and managed by the village government, and private forests owned and managed by individual landowners. However, minor subsistence activities are not bounded by these differences in land ownership and are conducted across the boundaries. Behind the non-existence of land-ownership boundaries for minor subsistence is the supportive notion that the *yama* landscape is a commons and that gathering must be practiced in accordance with local rules. Two rules are particularly important: "finders, keepers" and a customary obligation to share.

"Finders keepers" and the knowledge economy of mushroom sites

Mushroom gathering in the village is conducted in accordance with a simple rule: "finders, keepers." In other words, the person who discovers a mushroom is the one who can remove it from the forest. It seems likely that most people believe that gathering mushrooms is not a difficult task. However, mushrooming is in fact anything but easy. Blindly tramping through the steep forest slopes will give you a good workout, but it is unlikely to lead to any mushrooms. The reason that local inhabitants are able to rather easily locate mushrooms in the forest is that they know the locations of *shiro*, the sites where mushrooms grow in great abundance. Since mushrooms grow in mycelium, they appear annually in roughly the same sites. Accordingly, if one knows the location of *shiro* sites then one can efficiently and confidently gather mushrooms.

The fact that knowledge of *shiro* sites is vital to gathering mushrooms, and that the rule of "finders, keepers" is in play, creates competition in mushroom gathering (Figure 9.5). As long as the rule of "finders, keepers" is in play it is impossible

Figure 9.5 Gathering mushrooms at a *shiro* site (pre-earthquake: 2005).

for anyone to own *shiro* sites. *Shiro* are only known cognitively. When the *shiro* of wild mushrooms are discovered by a new individual then the use right for those mushrooms can potentially shift to that individual. For that reason, one characteristic of *shiro* is that they are subject to unstable use rights.

Villagers take a variety of measures to protect their use rights. They never place any sort of landmark at a *shiro* site but rather record the knowledge in their minds. They also never inform other villagers, or even other family members, of the location of *shiro* sites. In order to prevent anyone from following their tracks straight to *shiro* sites, villagers follow circuitous paths to reach these sites. Since *shiro* sites cannot be detected except when mushrooms are growing, villagers are careful not to convey to others the prime time for mushroom collecting. These are a few of the many obstacles that villagers put in place to protect their use rights.

A duty to share and the circulation of mushrooms through the village

There is one more highly important rule to be followed when gathering mushrooms. Gathered mushrooms are not to be monopolized but must instead be shared with other members of the community. Thus, while mushrooms gathered from the forest are for home consumption, it is rather more appropriate to say that mushrooms are passed between homes as gifts to friends and neighbors. The reason that mushrooms cannot be selfishly consumed is that there is a perception that the forests of the village are not individual property but rather the shared property of the community, or a commons. Additionally, since the *shiro* of the growing mushrooms cannot be seen, all mushrooms must be collected. Thus it is difficult to adjust the amount of mushrooms harvested. Some people note that they only eat a small portion at their home and that the majority of mushrooms gathered are given to others—for example, those who might not be able to collect mushrooms themselves. In this way mushrooms are passed around the village as gifts without any financial transactions.

As indicated by the above discussion, the fruits of minor subsistence activities are as a matter of course given to others and are thus "circulating resources."[3] This is entirely in contrast to rice cultivation and other products of major subsistence activities that are clearly the property of individual households. Since sharing among the community is seen as a matter of course, all of the inhabitants of Kawauchi have long looked forward to the mushroom-gathering season. As such, minor subsistence has been a means for smoothing social relations and binding the community together.

The continuing draw and degradation of minor subsistence practices

The forest areas of Kawauchi were contaminated as a result of the nuclear accident. Even worse, radioactive contamination continues to accumulate in the fruits of the forest, the plants, fish, and animal species that were formerly harvested and shared throughout the village. However, as problematized at the outset of this chapter, forests in the village have not been subject to any decontamination measures, nor has any compensation been provided for restricted use of these areas due to the narrow commercial calculus that defines both decontamination and compensation. The problem here is that minor subsistence practices are in fact one of the key features of rural agricultural life drawing residents back to the community. Evacuees wish to resume mushroom gathering and other minor subsistence practices but are faced with troubling dilemmas. Should they venture into contaminated forests to conduct minor subsistence practices? Should they continue to share the fruits of the forest? The next section explores these tensions.

The push and pull of minor subsistence practices for returning evacuees

While reasons for returning to the village vary greatly by individual, one that returnees often mentioned was a general unease with metropolitan life. Following the disaster, approximately one third of the population of Kawauchi evacuated outside the prefecture, with many of these individuals evacuating to the Tokyo metropolitan region in order to be close to children or relatives. Many of the rural villagers of Kawauchi noted major difficulties in adapting to the very different landscape and lifestyle of the metropolis: they were bored with metropolitan life, with the lack of physical activity and the lack of things to do, and they came to strongly desire to return to their lives in Kawauchi. Evacuees recalled their way of life in Kawauchi, particularly planting rice and conducting minor subsistence activities in the forests of the village. In other words, for many residents, minor subsistence practices were one of the key factors that drew them back to the village. By experiencing metropolitan life, they began to recognize the importance of their former ways of life.

Tragically, however, it is no longer possible to pursue minor subsistence in the *yama* landscape of the village light-heartedly. As described by Taira *et al.* (2014), products obtained from the forests of Kawauchi are highly contaminated (Table 9.2). A thorough analysis of the radioactive contamination of 2,564 food products indicated that 425 of these products were found to contain radioactive cesium over the national limit of 100 Bq/kg set for radiation in food. The degree of contamination is particularly high for products gathered from the forests of the village. Over 40 percent of wild *sansai* edibles and mushrooms were found to be over the limits for radiation in food, as were 64 percent of wild boar and pheasant and, although only a small sample size, 80 percent of all char and other fish. This contrasts with the fact that most agricultural products are within the limits for radiation in food. Wild *sansai* edibles and mushrooms were often found to be far over the limits, with one sample containing an incredible 9305Bq/kg.

Mushroom gathering after the disaster

Despite the difficult conditions currently confronting minor subsistence activities in the village, a number of individuals continue venturing into the forest. This section explores why these individuals continue to venture into the forest to gather mushrooms that may not be consumed.

Table 9.2 Radioactive contamination of foods in Kawauchi

	Vegetables		*Wild plants Mushrooms*		*Fruits*		*River fish*		*Meats (wild boar)*		*Crops*		*Others*		*Total*	
Area	n_S[1]	n_E[2]	n_S	n_E	n_S	n_E	n_S	n_E	n_S	n_E	n_S	n_E	n_S	n_E	n_S	n_E
1 Takadashima	346	1	179	63	41	2	3	3	17	12	12	0	23	7	621	88
2 Nishigo	222	0	41	9	31	0	0	0	6	2	5	0	14	1	319	12
3 Higashigo	150	0	67	31	32	0	0	0	15	11	5	0	12	3	281	45
4 Mochidome	174	2	100	26	40	1	1	1	4	3	8	0	15	3	342	36
5 Sakashiuchi	138	0	148	85	31	0	0	0	30	12	3	0	15	2	365	99
6 Nishiyama	215	1	95	36	53	1	2	2	28	25	14	0	18	6	425	71
7 Higashiyama	111	1	98	53	26	0	4	2	12	7	7	0	24	6	282	69
8 Modo	2	0	9	4	2	1	—	—	—	—	—	—	6	1	19	5
Total	1358	5	737	307	256	4	10	8	112	72	54	0	127	29	2654	425
Excess rate (%)[3]	0.004		**41.7**		0.016		**80.0**		**64.3**		0.0		22.8		16.0	
Maximum (Bq/kg)[4]	299		**9305**		163		**1440**		**5457**		30		2048		—	

Source: Author, based on Taira et al. (2014).

Notes

[1] n_S: number of samples.

[2] n_E: number of samples that exceeded the standard limit for radiocesium (100 Bq/kg for general foods).

[3] Excess rate: average excess rate of the standard limit for radiocesium for all areas of the village.

[4] Maximum: maximum value of radiocesium concentrations.

There are two types of individual who continue to venture into the forest in search of mushrooms despite contamination. The first group is relatively small in number, consisting of a few individuals who are aware of the high degree of contamination but continue to consume products gathered from the forest. These individuals note that these are foods they have gathered and consumed for years and that it would not be appropriate to leave these cherished foods. These individuals are elderly and unconcerned about the health effects of radiation due to their age. They would like to continue to consume the fruits of the forest for the remainder of their lives. However, even these individuals do not eat products from the forest as regularly as they did before the disaster. The second group is larger in number and consists of individuals who venture into the forest to collect mushrooms that will not be consumed in order to safeguard their *shiro* sites. Even if it is not presently possible to consume mushrooms, resuming gathering and consumption of mushrooms in the future necessitates safeguarding *shiro* sites. Villagers note that in order to prevent others from finding *shiro* they smother mushrooms that are growing at *shiro* sites. They state that it is difficult and painful to destroy mushrooms at *shiro* sites that they have long guarded and utilized.

As a result of radioactive contamination minor subsistence has become an activity that is limited only to those directly involved with it. Sharing foods gathered from the forest would inevitably lead to conflict. For example, even if a particular mushroom happens to be under the limits for radiation in food, it would be rather problematic to share this mushroom with a neighbor when it is known that mushrooms are generally contaminated.

Conclusion: the importance of the *yama* landscape

This chapter has examined the problems caused by the contamination of village forests for people who have returned to the evacuated areas. What I would first like to emphasize is that forests are an indispensable component of the area of life activity of a village for local residents. To be certain, from a purely economic calculus these forests are only important to a limited extent. However, villagers have long practiced minor subsistence in the forested spaces of the village.

In short, from the perspective of villagers, the contamination of forests has resulted in the serious contamination of products they would regularly consume. This has lowered the quality of life in the area. What I would like to emphasize here is the way in which a man-made disaster can result in serious contamination of the food supply, and not simply as a source of subsistence or income but as an important social good that enriches the quality of social life. This has been pointed out previously. Dyer (2002) uses the case of an oil-spill disaster in Alaska to show how resources for self-subsistence were devastated and how this resulted in diminished quality of life for residents. What has become clear in the course of this chapter is that despite serious contamination, residents continue to venture into the forest to conduct minor subsistence practices. This demonstrates how important these practices are to villagers. Indeed, these practices are so important that they have drawn some individuals back to the village.

What emerges from the analysis of this chapter are the issues and tasks ahead for recovery policies. Current decontamination policies are based on objectively assessing damages from the perspective of basic health; in other words, bare survival (Chapter 6). At the minimal level of bare existence the *yama* landscape is perhaps inessential. However, even at the level of bare existence, when we look at the way of life that the people in this village have practiced, it is clear that the forests of the village cannot and should not be overlooked. Four years after the disaster we have now reached a situation where it is necessary to move from emergency responses to a consideration of long-term recovery policies. Necessary in that effort is a policy debate based on the way of life of the people of the afflicted areas.

What types of measure should be adopted for the forest? Decontamination of village forests is difficult from both a technological and financial perspective. Additionally, decontamination of steep forest environments could potentially result in other disasters, including landslides and flooding. Faced with this difficult situation, the mayor of Kawauchi noted during an interview with the author that "we are aiming to continue managing the forest over time in order to renew it. I don't know how long it will take but we will renew it by cutting out the old contaminated trees and planting young ones."

The renewal of the forests will certainly be a long and difficult process. One of the tragic effects of radioactive contamination has been to sever the ties between villagers and village forests. Nevertheless, as demonstrated by this case study of Kawauchi Village, many people continue to venture into the forest. Some readers may see the continuation of minor subsistence practices in a negative light, as a source of potential risk and conflict. However, I believe that it is imperative to see the continuation of these practices as positive. Renewing the forests of the village is only possible if villagers continue to manage and utilize them. In sum, efforts to revitalize and conserve village forests pivot on the continuing presence of villagers who enjoy utilizing the forest and are willing and able to cultivate a passion for it among their neighbors and descendants.

Notes

1 This chapter uses a model of analysis called Life Environmentalism, a leading paradigm within environmental sociology in Japan. In wrestling with environmental problems, and the problem of pollution in particular, the field of environmental sociology in Japan has long pointed out the need to adopt the perspective of the victim. Life Environmentalism represents one of these perspectives. For more on this approach see Torigoe (2007, 2014).

2 The nine types of damages eligible for compensation are: 1) damages stemming from evacuation, 2) damages stemming from designation of dangerous to travel areas (for example, damages to business operations of transportation companies), 3) damages stemming from restrictions on agricultural, forests, and fisheries products, 4) damages stemming from government restrictions (for example, limits on water intake and water utilization in schools and other facilities), 5) reputational damage, 6) indirect damages (for example, evacuation of trading partners resulting in damages to business operations), 7) damages stemming from recovery operations, 8) damages stemming from voluntary evacuation and 9) damages stemming from decontamination.

3 Saito (2009: 165) also notes that mushrooms gathered from the *yama* landscape are actively shared with other villagers.

References

Dyer, Christopher L. 2002. "Punctuated Entropy as Disaster Induced Culture Change." In *Catastrophe and Culture*, edited by A. Oliver-Smith and S.M. Hoffman, 159–85. Santa Fe, NM: NM/SAR Press.

Fukuta, Ajio. 1989. *Jikan no Minzokugaku · Kukan no Minzokugaku* [Folklore Studies of Time and Space], Tokyo: Mokujisha. [In Japanese.]

Ikegami, Koichi. 2007. *Mura no Shigen wo Kenkyu Suru—Field Kara no Hasso* [Studying Village Resources: Thoughts from the Field], Tokyo: Nosangyoson Bunka Kyokai. [In Japanese.]

Masamura, Toshiyuki. 2013. "Higashi Nihon Daishinsai no Risk Mondai—Chi · Muchi · Ishi Kettei" [The Risk Problems of the Great East Japan Earthquake: Knowledge, Ignorance, Decision making]. *Japanese Sociological Review*, 64(3): 460–73. [In Japanese.]

Matsui, Takeshi. 1998. *Bunkagaku no Datsu Kochiku—Ryukyuko kara no Shiza* [Cultural Studies' Limits and Potentials: Perspectives from the Ryukyu Islands], Ginowan: Yoju Shorin. [In Japanese.]

—— 2004. "Minor Subsistence to Kankyo no Habitus-ka" [Acquire a Habitus by doing Minor Subsistence]. In *Ryukyu Retto—Shima no Shizen to Dento no Yukue* [The Ryukyu Archipelago: The Future of Island Nature and Traditions], edited by T. Matsui, 103–26. Tokyo: University of Tokyo Publishing. [In Japanese.]

Ministry of Economy, Trade and Industry. 2014. *Hinan Shiji Kuiki no Gainenzu* [Schematic Diagrams of Zone Evacuation Order], Tokyo: Ministry of Economy, Trade and Industry.

Ministry of Environment. 2013. *Josen Kankei Guidelines* [Decontamination Guidelines Second Edition], Tokyo: Ministry of Environment.

Saito, Haruo. 2009. "Hansaibai to Ruru—Kinoko to Tsukiau Saho" [Semi-Cultivation and Rules: Manners for Dealing with Mushrooms]. In *Hansaibai no Kankyo Syakaigaku* [Environmental Sociology of Semi-Cultivation], edited by Miyauchi, T., 155–79. Kyoto: Showado Publishing. [In Japanese.]

Seki, Reiko. 2012. *Keikai Kuiki Minaoshi ni Tomonau Nahara-cho Jyumin Tyosa* [Results of a Survey of Nahara-cho Residents on the Re-organization of the Restricted Areas]. Nahara-cho and Group for the Study of a Large-Scale and Compound Disaster. Accessed December 20, 2015. www2.rikkyo.ac.jp/web/reiko/lspcd/results/. [In Japanese.]

—— 2013. "Kyosei Sareta Hinan to Seikatsu no Fukko" [Nuclear Refugees and the "Reconstruction of Life and Living"]. *Journal of the Japanese Association for Environmental Sociology*, 19, 45–60. [In Japanese.]

Taira, Yasuyuki, Naomi Hayashida, Makiko Orita, Hitoshi Yamaguchi, Juichi Ide, Yuukou Endo, Shunichi Yamashita, and Noboru Takamura 2014. "Evaluation of Environmental Contamination and Estimated Exposure Doses after Residents Return Home in Kawauchi Village, Fukushima Prefecture." *Environmental Science & Technology*, 48 (8): 4556–63.

Torigoe, Hiroyuki. 2007. Land Ownership for the Preservation of Environment and Livelihood, Afrasian Working Papers, Volume 29.

—— 2014 "Life Environmentalism: A Model Developed under Environmental Degradation." *International Journal of Japanese Sociology*, 23(1): 21–31.

10 Refusing facile conclusions and continuing to tackle an aggregating disaster

Mitsuo Yamakawa and Daisaku Yamamoto

The earthquake and subsequent tsunami of March 11, 2011 triggered the meltdown and hydrogen explosion of multiple reactors at the TEPCO Fukushima Daiichi Nuclear Power Plant. The Great East Japan Earthquake Disaster took many lives and inflicted far-reaching physical, physiological, and economic damage. Yet these are not the only losses that we have had to endure. The disaster also eroded public confidence in expert knowledge and the kind of science on which the "nuclear village" and its power had been based. Reflecting on this situation, one of the underlying questions of this book has been what knowledge is truly useful in supporting reconstruction and redevelopment processes. While there are certainly no simple answers, the chapters here point towards certain shared understandings about how this difficult question can and must be answered. Real reconstruction and redevelopment, we believe, are not merely about restoring elements of the built environment such as roads and buildings; neither is "reconstruction and redevelopment support" solely about providing communities with a ready-made plan to be implemented. Rather than something imposed from above, reconstruction and redevelopment involve respecting and supporting the people who are willing and already acting to bring back their lives and livelihoods in their own ways. This is of course not to discount the importance of an "objective" empirical scientific understanding of spatially uneven radioactive contamination, nor critical inquiries into political structures and power relations that are responsible for various societal fissures. These are absolutely essential issues, and this book has explored a number of them. What we want to emphasize is our will to generate and use knowledge not for the sake of scholarship or criticism but for supporting those who are most in need. In this final chapter, we first articulate the cumulative nature of the nuclear disaster and then identify the many issues that remain unresolved five years after the initial accident. We then discuss a few theoretical issues before concluding.

The cumulative nature of the damages resulting from the nuclear disaster

As many of the previous chapters demonstrate, what makes the Fukushima nuclear disaster entirely different from an earthquake or tsunami disaster is the way in which radioactive contamination has rendered long-inhabited communities indefinitely uninhabitable, thereby significantly delaying and complicating any unified approach to recovery and reconstruction. If we conceptualize the damages from the nuclear disaster in somewhat schematic fashion, we see that three layers of issues accumulated one upon another. Further, since each of these issues remains unresolved, they have aggregated in a complex and vicious combination. Although complexly interwoven, it remains possible to untangle the processes whereby these three layers were aggregated, if we approach this process from the perspective of the successive phases of evacuation and evacuee life (Figure 10.1).

The first layer of damages was generated during the process of evacuation, during the trials and hardships of leaving behind a community to seek shelter in evacuation centers. In contrast to the damages from the earthquake or tsunami, the direct impacts of the nuclear disaster on human lives, such as death and injury, were relatively small, and beyond the nuclear plant, which was of course completely devastated, the destruction to buildings was also relatively minor.

Most damage to buildings resulted instead from the fact that forced evacuation prevented residents from performing maintenance on their homes and other properties, meaning that most damage was caused by leaks in the roof or from animals, including now feral livestock. In terms of the damage to farmland and farm products—both actual contamination and stigmatization—problems were

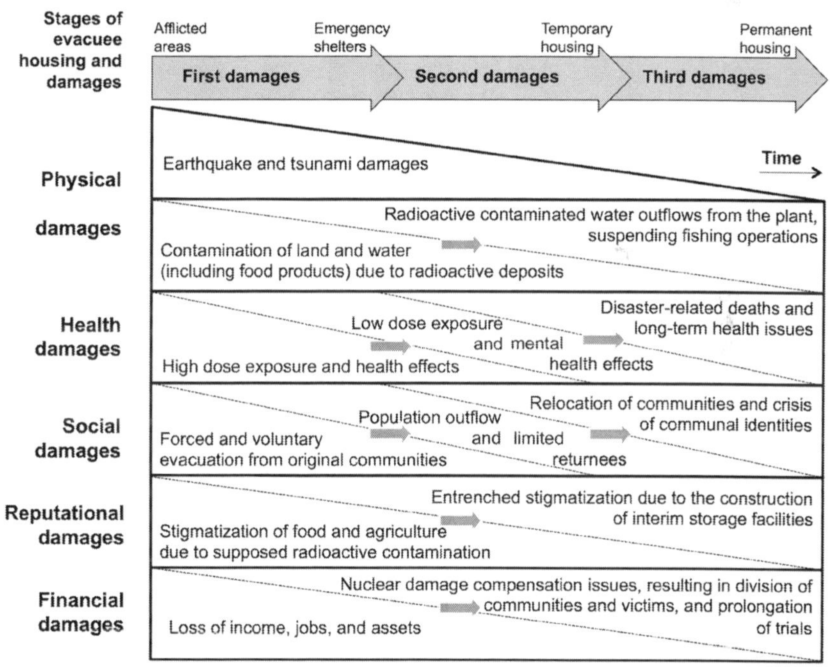

Figure 10.1 Accumulated damage from the nuclear accident.

Source: Authors, with the cooperation of Akira Takagi.

worsened by the fact that radiation was invisible to humans, that there were not enough measurement devices, and that the standards for shipment restriction fluctuated. Although direct and immediate health concerns about radiation exposure may have been limited to the technicians who were working on a cold shutdown to prevent further explosions in the reactor buildings, more indirect and widespread concerns among the inhabitants of the communities near the nuclear-power plant were the effects on life and health associated with the forced evacuation to avoid lower-level exposure. Many communities outside the official designated areas were also evacuated as individuals distrusted the provisional limits set for radiation exposure dose levels and in particular were highly concerned about the health of their children, leading them to pursue so-called voluntary evacuation beyond the prefecture (Chapters 4 and 5). These real and perceived differences regarding the effects of radiation resulted in uncoordinated and divisive evacuation practices that placed members of the same communities and even households into conflicting positions (Yamakawa 2013).

A second layer of damages emerged as evacuees transferred from evacuation centers to temporary housing, including prefabricated buildings and rentals (Figure 10.1). Due to the evacuation designations that followed the nuclear accident, evacuation life in shelters and temporary housing became indefinite and caused many evacuees to lose hope for return, recovery, or reconstruction. Not only is the soundproofing and climate control of temporary housing insufficient, but the cramped interior space is designed to accommodate nuclear families, forcing grandparents to live separately from their children and grandchildren. This had led to the emergence of numerous health issues among the elderly and also seems a likely factor behind the increase in the number of earthquake-related deaths. The number of individuals pursuing voluntary evacuation to avoid low-level radiation exposure for children has also significantly increased, leading to a great increase in the number of husbands and wives now living in distant separation. Many mothers who fled the area for the safety of their children are now concerned that they will be ostracized for having "fled" the community. The main concern of residents of the afflicted areas shifted from fears of radiation to concerns about their ability to make ends meet. The drawing up of evacuation boundaries further undermines the unity of local society due to the way in which it arbitrarily determines whether or not a household will receive nuclear-disaster compensation (Yamakawa 2014a).

A third layer of damages begins to appear as evacuees move from temporary housing to permanent housing, including their original homes and public-reconstruction housing (Figure 10.1). The most important prerequisite for the return of evacuees to the evacuated areas is the reduction of radiation dose levels. To that end national and local governments are providing massive funding for decontamination operations (Chapters 6 and 7). However, the scope and results of these efforts are limited. Most problematically, while all substances containing radioactive materials collected during decontamination will need to be transported to interim storage facilities, construction of these sites has been plagued by controversy and seriously delayed. The construction of interim storage facilities during a period in which final disposal sites have not been determined leads to further stigmatization and sends conflicting signals to those who wish to return to adjacent areas. In addition, although the issue of radioactive water leaking from the plant was present immediately after the accident, the source of this leak remains unidentified. Contamination of groundwater is increasing and, as this water eventually flows into the sea, this problem has led to suspensions of commercial fishing operations. Vast extents of forest also remain untouched by decontamination work (Chapter 9). The nuclear-accident compensation issue has shifted from the stage of consultation to that of legislation. The prolongation of trials not only negatively affects the health of nuclear evacuees but also further heightens the obstacles to recovery of life and livelihood.

In sum, the peculiar nature of the damages from the nuclear disaster is found precisely in their cumulative and complex layering. That is, while a primary layer of issues remained unresolved, the victims and afflicted areas were hit with a second and third wave of issues. It is this cumulative effect that has given rise to earthquake-related deaths and reduced evacuees' will and purpose (Yamakawa 2014b). Furthermore, designated evacuation areas have been divided into three distinct types depending on the levels of radioactive contamination—"difficult-to-return areas," "residence restricted," and "evacuation lift preparation"—in which different regulations and policies apply. Because each of these different areas may be found within the borders of many afflicted municipalities, and because local governments must decide on when to lift "evacuation lift preparation" area designations, the process of return, reconstruction and, recovery is highly complex and inherently political, and must be coordinated both within and between local governments.

Ongoing damages and unresolved issues

An unresolved accident: the discharge of contaminated water

Oceanic waters and groundwater around the plant are still being contaminated today as a result of the water leaking from the plant. The flow of highly contaminated water into the ocean was first confirmed on April 2, 2011. It was estimated that 4,700 Bq of radioactive materials flowed out from the plant site from April 1–6 that year.[1] Further, the extent of the damage to the reactor containment vessel remains undetermined, and water used to cool melted fuel during the accident has mixed with the 300 m^2 of water that flows out daily, resulting in continuing increases in the amount of contaminated water inside the reactor building. Although this contaminated water continues to be processed by the multi-nuclide removal facility (or Advanced Liquid Processing System (ALPS)), a large amount of contaminated water is being stored in tanks within the site of reactor unit one (TEPCO 2015).

The national government and TEPCO have developed three basic principles as well as a concrete framework for dealing with the contaminated water. The first principle stipulates the removal of the sources of contamination. This entails efforts to remove the highly contaminated water under the reactor building and in the trench outside the building and to decontaminate this water using ALPS. The second principle stipulates that water should be prevented from coming near the contamination source. Such efforts include concrete measures such as paving on site to prevent precipitation from seeping into the ground, and the installation of bypass pipes, pumped wells, and frozen-soil shielding walls to ensure that influx groundwater does not mix with the cooling water in the damaged reactors. The third principle is to prevent water from leaking from the site. To that end a number of measures have been implemented, including the use of soluble glass (sodium silicate) in the ground, sheet piles on the ocean side of the plant to block water paths, and the construction of more and better-designed contaminated water tanks (TEPCO 2015).

Implementation of these contaminated-water countermeasures has reduced the concentration of radioactive materials in the ocean waters around Fukushima Daiichi from 10 to 1 parts per million. However, it is not the case that the leakage of contaminated materials has been completely stopped. Moreover, decontamination has not been implemented within the harbor, and the coating of the seafloor in a special cement mix has only curbed the diffusion of radioactive materials in mud and sand (TEPCO 2015). Finally, radioactive materials deposited in the inland mountainous regions of the prefecture have adhered to fallen leaves and soil, and these radioactive materials are carried by precipitation into rivers and eventually to the ocean floor. While the degree of contamination is of course diluted and diffused throughout the massive volume of oceanic waters, investigations of water and marine soil contamination are lacking and detailed contamination maps have not yet been produced.

This outflow of radioactive contaminated water into the marine environment has resulted in shutdowns and other restrictions and has caused severe economic damage to the fishing industry of Fukushima Prefecture. The articulation of the perplexing issues of contamination together with a host of other factors has completely halted the fishing industry in the area. These other factors include: loss of fishermen—the basis of the industry—as a result of the massive tsunami, destruction of the infrastructure of life and livelihood such as homes and offices, destruction of fishing boats and other means of production, and destruction of shared infrastructure such as fishing ports (Hamada 2015). At present, along with the gradual lifting of shipping restrictions for aquatic products, the Fukushima fishing industry has strongly advocated for the restart of experimental testing with an eye towards resuming operations. Additionally, as a result of the redevelopment of nearby fishing ports and compensation received for nuclear-disaster damage, the reconstruction of fishing vessels has begun to progress. However, since the level of operations remains well below pre-disaster levels, fishermen have had their incentives reduced and, further, many fisherman have shifted into the tertiary sector. Since safety and working conditions are often better in this sector than in fishing, it becomes increasingly difficult to secure laborers to restart fishery operations.

Tasks ahead for radioactivity decontamination operations

The national government's response to this nuclear accident has been to advance radioactivity-decontamination operations at breakneck speed. Within the designated evacuation areas these operations are under the direct control and implementation of the national government. As of July 2015, decontamination was completed in all planned areas in four out of the eleven evacuated municipalities. In three other municipalities, residential areas were decontaminated by that date. Since these operations are implemented first in areas with relatively lower air dose rates, decontamination operations have been significantly delayed in municipalities with large difficult-to-return and limited-residence evacuation areas. Outside the designated evacuation areas, local municipal governments are in charge of implementing decontamination measures. In June 2015, the rates of progress for decontamination in these 36 municipalities were as follows: 90 percent of public facilities, 60 percent of residential areas, and 30 percent of roads (Reconstruction Agency 2015a).

"Decontamination" operations essentially entail the relocation and confinement of radioactive materials. Accordingly, securing interim storage facilities for the soil and organic materials containing radioactive materials is a prerequisite for advancing these operations. However, the reality is that interim storage-facility development is lagging far behind decontamination, and the majority of decontaminated materials have been stored at municipalities' temporary sites or "pre-temporary" sites in front of residences. These decontaminated materials are simply packed by the cubic meter into large woven plastic containers and left to sit under the blue sky. Over the course of four years these containers have begun to deteriorate and are susceptible to tearing during heavy rains and flooding resulting from typhoons, creating the highly problematic potential for decontaminated materials to leak. In addition, due to the decomposition of plants and other organic matter in the containers, it is imperative to conduct aeration operations for methane and other gases.

In October 2011, the national government indicated its basic guidelines for interim storage facilities and subsequently communicated these guidelines to the mayors of municipalities in Fukushima Prefecture. The guidelines contain four main stipulations: 1) the siting and maintenance of facilities will be conducted by the national government; 2) the national government will exert maximum effort to reach the goal of opening these facilities within three years after storage at temporary sites commences (i.e. January 2015); 3) only soil and waste from within Fukushima Prefecture will be stored; and 4) within 30 years of opening, interim storage materials will be placed in final disposal outside Fukushima Prefecture. Field surveys were conducted in April 2013. In February 2015, Fukushima Prefecture, Okuma Town, and Futaba Town agreed to accept transport to interim storage facilities and construction began. However, the sticking point of locating final disposal sites outside Fukushima Prefecture remains undecided, and there is fear that interim storage facilities will become de facto final disposal sites.

The unresolved issue of stigmatization of agricultural products

The stigmatization (*fuhyo higai*: reputational damage) issues affecting Fukushima agricultural products as a result of the nuclear disaster remain unresolved. The Dispute Reconciliation Committee for Nuclear Damage Compensation established by the national government has defined reputational damage as:

> damages resulting from the deferral or halting of the purchase of goods or services by consumers or trading partners due to fears of such goods or services being contaminated by radioactive materials as a result of information widely conveyed by the media.
>
> (Dispute Reconciliation Committee for Nuclear Damage Compensation 2011, author's translation)

In regard to agriculture, forestry, and fishing, as well as severely stigmatized food products, methods have been stipulated for specifying which goods and producing prefectures will be eligible for compensation. These apply "not only to a single product over the provisional limit for radiation in food where a shipping restriction was issued by the government but to all of similar types of products (i.e. agriculture, livestock, fisheries) from within the same area," but

> even to areas outside the area where the restriction was stipulated, for certain regions with similar characteristics (in particular areas in close proximity to where this accident occurred, and areas with geographical connections to the area where the restriction was issued), and also with similar distribution conditions (in particular geographical origin).
>
> (Dispute Reconciliation Committee 2011: 46, author's translation)

JA Fukushima, the regional agricultural cooperative, has been the point of contact for agricultural compensation issues in Fukushima. As of July 22, 2015, 208 billion yen has been charged to TEPCO and 205.4 billion yen has been received (JA Fukushima Chuo Kai 2015).

One might look at the situation described above and assume that compensation for damages to agriculture is proceeding apace. However, the Science Council of Japan points out in its "Urgent Recommendations" that damages to agriculture in Fukushima from the nuclear disaster

> are not limited to the damages to 'flow' that prevented radioactive contaminated products from being sold but also include destruction of agricultural land, the basic 'stock' of production, and the agricultural community, or the social overhead capital that maintains agriculture, as well as damages to intangible assets such as human resources and local brand recognition.
>
> (Science Council of Japan 2013: 3, author's translation)

Of the different types of damage, those to flow are relatively easy to calculate and, as noted above, the criteria for such damages have been stipulated in the "interim guidelines."[2] Damages to stock—the agricultural community, human resources, and brand value—are by contrast extremely difficult to calculate.

The Consumer Agency has conducted a survey of consumer consciousness about food and radioactive contamination every six months since February 2013 (Consumer Agency, Team of Facilitating Consumer Understanding 2015). According to the most recent surveys, 42–51 percent of respondents indicated that they want to eat foods with the lowest levels of radioactivity even if they are under the government-set limit for radiation in food. These rates have not decreased in recent surveys, indicating persisting consumer anxiety concerning radioactive contamination of food, arguably to a level beyond that which scientific assessments of risk would consider reasonable. Furthermore, 15–20 percent of respondents continue to state that they hesitate to purchase Fukushima products, while only 4–6 percent of respondents indicate that they hesitate to purchase products from the greater northeast region. This indicates, perhaps not surprisingly, that the stigmatization of food is strongly associated with the name "Fukushima," even though some areas of Fukushima are minimally affected by radioactive contamination and there have been radiation hot spots outside the prefecture.

Ryota Koyama and the Science Council of Japan have proposed a four-step inspection framework for overcoming the challenges of stigmatization. Stated briefly, the first step entails the creation of field-by-field maps of the distribution of radiation dose rates, while in an interlinked second step these maps are used to create databases pertaining to transfer rates for regions and products and to guide absorption countermeasures. Then, through the proper combination of the third step—the expansion of pre-shipment testing—and the fourth step—testing at sites of consumption—it will become possible to use the fulfillment of safety on the producers' side as a basis for guaranteeing "security" on the consumer side (Koyama 2012; Koyama and Komatsu 2013).

Disparities in compensation: forced evacuation and voluntary evacuation

Voluntary evacuees are often far more economically challenged than forced evacuees. A survey of evacuees from Fukushima Prefecture conducted in February 2015 found that both forced and voluntary evacuees identified concerns for physical health as the number one source of anxiety in daily life (Fukushima Prefecture 2015a). However, while forced evacuees identified a lack of plans for the future as their second greatest concern, voluntary evacuees identified concerns over money and expenditures as theirs. This difference is perhaps best explained by the fact that the percentage of voluntary-evacuee households in which only a few household members have evacuated is very high, indicating that they are charged with the burden of leading life at two locations (for example, bearing the costs of commuting and rental housing on their own) (Chapter 5).

In principle, compensation for the effects of the nuclear accident is for forced evacuees and, in addition, this compensation is paid out by TEPCO rather than the national government (Chapter 7).[3] Compensation for forced evacuees can be divided broadly into five categories: compensation for property (housing and land), compensation for damages related to securing housing in the process of evacuation and return, compensation for household valuables, compensation for suffering, and compensation for lost sales and disability.

As of May 8, 2015, the total amount of compensation paid out for damages was 914.7 billion yen.[4] According to Fukushima Prefecture, the amount of compensation paid out to a forced evacuee household of four is estimated to be 153.18 million yen for difficult-to-return areas, 150.3 million yen for limited-residence areas and 135.1 million yen for evacuation-lift-preparation areas.[5] In contrast, in 2011, compensation for voluntary evacuees was limited to individuals living within specifically designated voluntary-evacuation areas,[6] and this compensation only amounted to 400,000 yen per month for individuals under 18 and pregnant women and 80,000 yen per month for others. In 2012, 40,000 yen was added to the individual compensation figures, but clearly there remained a world of difference between compensation for voluntary and forced evacuees. In cases where voluntary evacuation is pursued, reasons for conferring compensation include increased living expenditures stemming from voluntary evacuation, damages for emotional stress resulting from a considerably inhibited ability to continue normal daily life, and expenses resulting from evacuation or return. However, since this amount of money is far from sufficient to support life at two sites, it essentially forces voluntary evacuees to return to an evacuated site or to pursue long-term residence at an evacuation site. To avoid this, there is a need to establish the right to remain in evacuation (Science Council of Japan 2014).

Between detached-residence orientations and the prolongation of rental living

Securing housing is an aspect of the reconstruction of evacuees' lives that cannot be overlooked or overemphasized. Compensation for earthquake-related damages was a relatively small percentage of the total compensation paid out in the nuclear-disaster-afflicted areas. However, since it was not possible to enter the designated evacuation areas, the majority of compensation for earthquake-related damages is actually for rapidly progressing deterioration due to the inability to perform maintenance. Such damage and deterioration are caused by mice, mildew, and heavy rains. Using a survey conducted by the Reconstruction Agency, we can see that while securing housing has certainly progressed to a certain level, between 2011 and 2014 homeownership grew from only 4 percent to 19 percent.

However, if we look at employment or finances for rebuilding life and livelihood, we find that employment-status figures dropped from 67 percent before the disaster (i.e. March 2011) to 40 percent in 2013, and that the unemployment rate shifted from 22 percent to 50 percent (Reconstruction Agency 2014). Among those in employment, the percentage of self-employed individuals or individuals working in companies where operations were not suspended or have been restarted dropped from 18 percent to 5 percent, while employment in private offices fell from 35 percent to 24 percent. Part-time employment dropped from

7 percent to 5 percent. The drop among those employed as civil servants or in organizations was only from 7 percent to 6 percent.

What can be seen from the above data is that there is a growing polarization between evacuees who desire to live in single-family residences and will return to the evacuated areas and those who will have economic difficulties in returning. Over half of evacuees would like to stay where they are living now, and the number who would like to see "an extension of the period for living in emergency rental units" is also increasing. The reason for this is not limited to "evacuation orders" or "concerns over radiation," but also includes economic concerns identified by over 40 percent of evacuees, such as "not having reconstructed housing" or "concerns over living expenses."

Nuclear-accident-related deaths and physical and mental-health problems

One unique and highly unfortunate aspect of the damages resulting from the nuclear accident is the high number of disaster-related deaths. The number and spatial distribution of direct fatalities from the Great East Japan Earthquake Disaster were reflective of the size of the massive tsunami and thus were the greatest in Miyagi Prefecture (9,621), followed by Iwate Prefecture (4,672). In these prefectures, disaster-related deaths numbered 450 and 909 respectively, or around 10 percent of the total number of direct fatalities. In contrast, while the number of direct fatalities in Fukushima Prefecture was relatively low, at 1,603, disaster-related deaths numbered 1,867, more common than direct fatalities (Reconstruction Agency 2015b). The Reconstruction Agency's report shows that a greater number of deaths of Fukushima residents than those of Iwate and Miyagi residents were caused by the physical and mental fatigue of moving to and living in shelters, indicating one distinctive effect of the nuclear accident (Reconstruction Agency 2012).

Life in evacuation shelters and temporary housing has been going on now for five years. According to Fukushima Prefecture, two thirds of households in evacuation have noted that at least one member of the household has suffered from physical and mental disorders. Similarly, when asked about concerns and difficulties with current living arrangements, 63 percent of evacuees noted problems with their health or the health of family members. Here we can see the way in which life in temporary housing has worn down the physical and mental health of evacuees. Perhaps the biggest problem is that 10 percent of evacuees noted that they have no one with whom to discuss their plight (Fukushima Prefecture 2015a).

To grasp the state of residents' physical and mental health—burdened as they were with emotional distress and trauma resulting from radiation concerns, the trials of evacuation, and the loss of personal property—and provide proper care, Fukushima Prefecture implemented from fiscal year 2011 a survey on health and living habits. The results indicate that, for children, the indicators of health and exercise have all improved with the passage of time but are still below national averages. For adults, too, the results indicate that overall mental health and responses to trauma have improved with the passage of time but are still far below national averages. This is particularly the case for women and older individuals. Additionally, while the daily habits of adults have improved, many are facing problems with sleeping, drinking, smoking, and exercise (Fukushima Medical University 2015).

Over the same period, Fukushima Prefecture has surveyed its citizens' health in order to understand the effects of radiation exposure. However, the screening rates for the survey have declined, from 31 percent to 22 percent among individuals over 16 years old, and from 65 percent to 36 percent among children 15 and under. The low and declining rates are alarming. Fukushima Prefecture speculates that the declining participation rates are due carelessness and the short period in which examinations were available, but Hino (2013) argues that the low rate of examination actually represents a lack of determined effort by public officials.

Open challenge for theoretical development

In short, the hard empirical lessons that we have learned, and continue to learn, from this nuclear disaster are not only the severity of each biophysical and socioeconomic damages, but how these damages have accumulated and interacted in the areas contaminated by radioactive materials and in the lives of the people who once resided in these areas. In part due to the proximity to and immediacy of the affliction, many of the contributors to this volume, especially those who are based in Fukushima, focus on documenting the nature and extent of the damages and do not articulate theoretical ideas and implications in the ways that many English-speaking readers would expect. To be clear, our primary goal is not to fit the unfolding realities into the mold of existing theoretical language or abstract conceptions, but we concur that a disciplined but more assertive use of theoretical language may further facilitate intellectual conversation across different academic cultures and help us articulate commonalities and differences across disaster experiences in other parts of the world. Those are the tasks left for future undertakings.

Nevertheless, we can already identify based on the contributions to this volume a few parameters for future theoretical development, which admittedly reflect our (and many contributing authors') disciplinary background, i.e. geography. First, any theoretical undertakings, let alone policy-oriented actions, must recognize that materialities matter hugely and fundamentally at different geographical scales (Massey 2005), cautioning us against, for example, discussing the rights and wrongs of nuclear power or redevelopment strategies in the abstract. At the continental scale, as Nakamura (Chapter 1) makes us realize, a simple macro-scale geophysical reality is that the country has built more than 50 nuclear reactors along some of the most seismically active regions of the entire globe.[7] The diffusion of radioactive materials within the country is thoroughly influenced by geomorphological, soil, and land-use conditions, so that one cannot easily infer its dynamics from the limited number of other nuclear-disaster experiences (Chapter 2). At the micro biophysical scale we face the fact that radioactive materials differentially affect our bodies depending on age, a fact that has clear significance for the perceptions and spatial behaviors of evacuees (Chapters 4 and 8).

Another geographic theme that has emerged is the role of maps and geographic information, but this is far from a simple matter of producing accurate maps. To be sure, there is little question of how essential maps are in any disaster situation. In this nuclear disaster, too, detailed maps and other geographic data of invisible radioactive contamination have been some of the most useful information for the public. Yet because of the apparent objectivity and naturalness of geographic/cartographic information, it can surreptitiously but forcefully facilitate the exercise of power (Sheppard 1995). Oda (Chapter 3) illustrates this problem in the context of evacuation, in which poor communication of geographic representation of "risky areas" amplified disaster damage and stigmatization. Grassroots-level measuring and mapping of radioactive micro hotspots have been critical in informing the pubic (Chapters 6 and 7), echoing the growing importance of volunteered geographic information (VGI) or the "wikification" of GIS (Goodchild 2007; Sui 2008). Yet we have also encountered situations such as grassroots organizations themselves hesitating to disclose detailed radiation-hotspot measurements of farmland in Fukushima City when it became apparent that the information might create tension among farmers. Accordingly, closer analyses of these realities have much to contribute to the literature on critical GIS (for example, Elwood 2009).

In a similar vein, the disaster has made it clear that drawing the geographical boundaries of different evacuation zones is not simply a technical matter of presenting "objective" information; rather, it is a fundamentally political process in which someone decides what information is to be measured, at what scale and resolution, how it is visualized, what levels of radiation are considered harmful, whether or not to release the information (i.e. the case of SPEEDI), and so on (Chapters 6 and 7). The defined boundaries then generate various social dynamics (for example, compensation disparities and conflicts among neighbors)—a vivid example of what Anglo-American geographers have long called a socio-spatial dialectic process (Soja 1980) or of the tension between "conceived space" and "lived space" (Lefebvre 1991).

Finally, the importance of place specificity and place-based understanding has been a repeated theme in the volume. Unlike many other disasters that the country has experienced, in the Fukushima nuclear disaster there are no broadly agreed standards for "safe" conditions to return because of the nature of radioactive contamination of the environment. In order to understand why evacuees return, wish to return, or do not want to return, we must understand not only the physical conditions (i.e. radiation levels and infrastructural reconstruction) of their hometowns, but we must also understand household livelihood strategies, social values of place/landscape, and local customs and traditions rooted in place (Chapters 4, 5, and 9). In this regard, theoretical perspectives that emphasize the importance of local livelihood strategies, such as the sustainable-livelihood literature (for example, Chambers and Conway 1991; Lu and Lora-Wainwright 2014), may provide a useful lens through which the situation and prospects for redevelopment of the afflicted communities can be understood. We would also like to insist that valuable theoretical perspectives and insights do not have to come from the English-language literature. In particular, Japanese folk-culture studies (*minzoku-gaku*) have long contributed to a critical and reflective understanding of local life-ways and development, especially of the rural countryside, and to inform other fields of social science in Japan, such as environmental sociology, as indicated by Kaneko (Chapter 9). Yet the literature has only recently made headway into the English-language literature (for example, Torigoe 2014), and there is much room for intellectual fertilization.

In short, a lot of tasks are untouched in terms of articulating empirical accounts in more refined theoretical language. We hope that this book provides a foundation for a fertile exchange of ideas and insights among those who have been observing and thinking about Fukushima *in* Fukushima, and those who are equipped with theoretical language that will help understand the empirical realities even further. Nevertheless, we wish to emphatically underscore what we consider the role of theories and theoretical debate should be: that is, to support those who are in the most need amid this continuing disaster. To that end, we concur with Torigoe and Kada, who assert:

> After all, "theories" are not some kind of "objective" beings that arise from the collection of facts; rather, they are merely analytical frameworks to interpret social phenomenon. That is, they are fictions. An instance when a theory, as a fiction, performs some useful functions is when the society determines that the theory has reality.
>
> (Torigoe and Kada 1984: 338, authors' translation)

Interestingly, they go on to say, "the criterion for assessing reality is not data, but is the sense of everyday life" (Torigoe and Kada 1984: 338, authors' translation). In other words, the validity of a theory should be judged not by the accuracy of collected data, nor by the level of sophistication or the logical consistency of the theory; rather, it should be judged by how it resonates with the intuitive sense (or consciousness) of those who are "explained," while also having the potential to create new subjectivities (Gibson-Graham 2006). Such a theory would likely take a form of "weak theory" that "(refuses) to extend explanation too widely or deeply, refusing to know too much" and "welcomes surprise, tolerates coexistence, and cares for the new, providing a welcoming environment for the objects of our thought" (Gibson-Graham 2008: 619). Such theorization for the redevelopment of Fukushima has barely begun.

Conclusions

The forward-focused Japanese government is eager to claim "reconstruction complete" at the 2020 Tokyo Olympics, and domestic mass media appear to be dancing to the same tune. Although it is true that there has been considerable progress in reconstruction over the past five years, a number of issues remain far from resolved, ranging from radioactive contamination to decontamination projects, stigmatization, compensation, housing, and the health of evacuees. None of these issues has been resolved. Instead, each has connected with the others in complex ways. We wish to emphasize, however, that it would be incorrect and irresponsible to conclude that Fukushima is forever damaged and unrecoverable. Just as the nuclear disaster was ultimately a human-induced disaster built upon the history of Japan's energy policies and industry practices (under the façade of peaceful use of nuclear power), these damages have been shaped significantly by decisions and policies implemented through the course of reconstruction. Therefore, rather than treating this disaster as the "tragedy of Fukushima" and fixating our understanding on that image, we must critically engage in a policy process for reconstruction and redevelopment both at the micro level (for example, specific reconstruction housing policies in specific localities) and the macro level (for example, a broader energy-security policy). What is needed more than ever is a critical mindset that refuses facile conclusions, as well as vigilant eyes to observe Fukushima for years to come.

Notes

1 As of May 28, 2015, approximately 442,000 tons of contaminated water containing tritium and 185,000 tons of contaminated water containing reduced concentration strontium were being stored on site (Nuclear Emergency Response Headquarters 2011)

2 Yet, as fruits and vegetables are subject to market processes of substitution or competition with other similar products, products from other areas, and processed foods, Abe (2013) and Hangui (2013) note that using market prices to identify reputational damages is not a simple endeavor. In extreme cases, some commentators even question whether reputational damage even exists. What is required is a multifaceted analysis not only of consumer behavior but of the power of local brands as reflected in the actions of buyers and sellers in wholesale markets, product placement on supermarket shelves, and the use of Fukushima products in school lunches (Yamakawa *et al.* 2014).

3 This is based on the Interim Guidelines (August 5, 2011) and the Interim Guidelines Fourth Supplement (December 26, 2013) of the Dispute Reconciliation Committee for Nuclear Damage Compensation (Ministry of Education, Culture, Sports, Science and Technology 2016).

4 The total compensation breakdown is as follows: 353.3 billion yen for voluntary evacuees, 1386.5 billion yen for businesses, 828.5 billion yen for organizations and local governments, 194.5 billion yen for individuals, and 151.9 billion yen in provisional payments.

5 The reason that compensation is larger for difficult-to-return zones is, first, that individual residents receive 7 million yen as compensation for loss of hometown and, second, because compensation to individuals (i.e. monthly payments of 100,000 yen) as well as for residential land, buildings, farmland, and forests increases along with the extension of the evacuation period (Fukushima Prefecture 2015b).

6 These are outside the evacuation-designated areas but still within roughly 30 to 80 km of the Fukushima Daiichi Nuclear Plant, and include Fukushima City, Koriyama City, and Iwaki City.

7 Indeed, there is still latent suspicion that some of the critical damage to the Fukushima Daiichi Nuclear Power Plant was caused by the earthquake itself, not by the tsunami, which would potentially require a much more comprehensive revision of the country's nuclear strategy. TEPCO appears very much unwilling to open up and engage in this debate.

References

Abe, Shiro. 2013. "Nosanbutsu Suisanbutsu kara Miru Fuhyo Higai" [Stigmatization Seen in Agricultural and Aquatic Products]. *Journal of Social Capital Studies* 4: 23–43. [In Japanese.]

Chambers, Robert and Gordon Conway. 1991. "Sustainable Rural Livelihoods: Practical Concepts for the 21st Century." IDS Discussion Paper 296.

Consumer Agency, Team of Facilitating Consumer Understanding. 2015. *Fuhyohigai ni kansuru Shohisha Ishiki no Jittai Chosa (No. 5): Shokuhin Chu no Hoshasei Busshitsu Nado ni kansuru Ishiki Chosa Kekka* [Fifth Survey of Consumer Awareness related to Stigmatization: Results of the Survey on the Awareness related to Radioactive Materials in Food], Tokyo: Consumer Agency. [In Japanese.]

Dispute Reconciliation Committee for Nuclear Damage Compensation. 2011. *Tokyo Denryoku Kabushi Gaisha Fukushima Daiichi, Daini Genshiryoku Hatsudensho Jiko niyoru Genshiryoku Higai no Hanni no Hanteinado ni kansuru Chukan Shishin* [Interim Guidelines for Determining the Extent of Nuclear Damage from the TEPCO Fukushima Daiichi and Daini Nuclear Power Plant Accidents]. Accessed March 14, 2016. www.mext.go.jp/b_menu/shingi/chousa/kaihatu/016/houkoku/__icsFiles/afieldfile/2011/08/17/1309452_1_2.pdf. [In Japanese.]

Elwood, Sarah. 2009. "Thinking Outside the Box: Engaging Critical Geographic Information Systems Theory, Practice and Politics in Human Geography." *Geography Compass*, 4(1): 45–60.

Fukushima Medical University. 2015. *Heisei 25 Nendo Kenmin Kenko Chosa* [Health Survey of Prefectural Residents, 2013 Fiscal Year]. Accessed March 14, 2016. www.pref.fukushima.lg.jp/uploaded/attachment/115330.pdf. [In Japanese.]

Fukushima Prefecture. 2015a. *Fukushima Ken Hinansha Iko Chosa Kekka (Gaiyoban: 2015 Nen 2 Gatsu Chosa)* [Results of the Survey on the Views of the Evacuees from Fukushima Prefecture (Summary: Study Conducted in February 2015)]. Accessed March 14, 2016. www.pref.fukushima.lg.jp/uploaded/attachment/113135.pdf. [In Japanese.]

—— 2015b. *Fukushima no Fukko ni Muketa Torikumi* [Efforts toward the Reconstruction of Fukushima]. Accessed March 14, 2015. www.pref.fukushima.lg.jp/uploaded/attachment/113135.pdf. [In Japanese.]

Gibson-Graham, J. K. 2006. *A Postcapitalist Politics*, Minneapolis: University of Minnesota Press.

—— 2008. "Diverse Economies: Performative Practices for 'Other Worlds'." *Progress in Human Geography*, 32(5): 613–32.

Goodchild, Michael. F. 2007. "Citizens as Sensors: The World of Volunteered Geography." *GeoJournal*, 69(4): 211–21.

Hamada, Takeshi. 2015. "Kaiyo Osen kara no Gyogyo Fukko" [Recovery of Fishery from the Aquatic Contamination]. In *Fukushima ni Roringyo wo Torimodosu* [Bring Back Agriculture and Fishery to Fukushima], by Takeshi Hamada, Ryota Koyama, and Masahiro Hayajiri, 215–303. Tokyo: Misuzu Shobo. [In Japanese.]

Hangui, Shinichi. 2013. "Severe Damage to the Agricultural Economy of Fukushima Prefecture as a Result of the Great East Japan Earthquake and Nuclear Disaster: Consequences to the Shipment and Distribution Stages of Fruits and Vegetables in the Year 2011." *Fukushima Ken Nogyo Sogo Henkyu Senta Kenkyu Hokokusho*, 126–9. Accessed March 14, 2016. www.pref.fukushima.lg.jp/w4/nougyou-centre/kenkyu_houkoku/radio_sp_2014/2014_02_35.pdf. [In Japanese.]

Hino, Kosuke. 2013. *Fukushima Genpatsu Jiko Kenmin Kenko Kanri Chosa no Yami* [The Darkside of the Fukushima Nuclear Accident Health Management Survey], Tokyo: Iwanami Shoten. [In Japanese.]

JA Fukushima Chuo Kai. 2015. "JA Gurupu Tokyo Denryoku Genpatsu Jiko Nochiku Sanbutsu Songai Baisho Taisaku Fukushima Ken Kyogikai no Torikumi nitsuite." Accessed March 14, 2015. www.acc.go.jp/jicst/NC/senmon/songai/siryo03/siryo3-3.pdf. [In Japanese.]

Koyama, Ryota ed. 2012. *Hoshano Osen kara Shoku to Nou no Saisei wo* [Recovering Food and Agriculture from Radioactive Contamination], Tokyo: Ieno Hikari Kyokai. [In Japanese.]

Koyama, Ryota and Tomomi Komatsu eds. 2013. *Nou no Saisei to Shoku no Anzen: Genpatsu Jiko to Fukushima no Ninen* [Recovery of Agriculture and Safety of Food: Two years of the Nuclear Accident and Fukushima], Tokyo: Shin Nihon Shuppan Sha. [In Japanese.]

Lefebvre, Henri. 1991. *The Production of Space*, London: Blackwell Publishing.

Lu, Jinxia and Anna Lora-Wainwright. 2014. "Historicizing Sustainable Livelihoods: A Pathways Approach to Lead Mining in Rural Central China." *World Development*, 62: 189–200.

Massey, Doreen. 2005. *For Space*. London: Sage Publications.

Ministry of Education, Culture, Sports, Science and Technology. 2016. *Genshiryoku Songai Baisho Funso Shingi Kai* [Dispute Reconciliation Committee for Nuclear Damage Compensation]. Accessed March 14, 2016. www.mext.go.jp/b_menu/shingi/chousa/kaihatu/016/. [In Japanese.]

Nuclear Emergency Response Headquarters. 2011. *Genshiryoku Anzen ni kansuru IAEA Kakuryo Kaigi ni taisuru Nihonkoku Seifu no Hokokusho* [Report of the Japanese Government for the IAEA Ministerial Meeting on Nuclear Safety]. Accessed March 14, 2016. www.kantei.go.jp/jp/topics/2011/pdf/houkokusyo_full.pdf. [In Japanese.]

Reconstruction Agency. 2012. "Higashi Nihon Daishinsai niokeru Shinsai Kanreshi nikansuru hokoku" [Report on Disaster-related Deaths from the Great East Japan Earthquake]. Accessed March 14, 2016. www.reconstruction.go.jp/topics/20120821_shinsaikanrenshihoukoku.pdf. [In Japanese.]

Reconstruction Agency. 2014. "Genshiryoku Hisai Jichitai niokeru Jumin Iko Chosa" [Survey of Residents of the Nuclear Disaster-Afflicted Municipalities]. Accessed March 14, 2016. www.reconstruction.go.jp/topics/main-cat1/sub-cat1-4/ikoucyousa/. [In Japanese.]

Reconstruction Agency. 2015a. "Fukko no Genjo to Kadai" [Current Situation and Problems in Reconstruction]. Accessed March 14, 2016. www.reconstruction.go.jp/topics/main-cat1/sub-cat1-1/150911_gennjyoutokadai.pdf. [In Japanese.]

Reconstruction Agency. 2015b. "Higashi Nihon Daishinsai niokeru Shinsai Kanrenshi no Shishasu" [The Number of Earthquake Related Deaths from the Great East Japan Earthquake (as of March 31, 2015)]. Accessed January 27, 2016. www.reconstruction.go.jp/topics/main-cat2/sub-cat2-6/20150630_kanrenshi.pdf. [In Japanese.]

Science Council of Japan, Fukushima Reconstruction Support Division of the Great East Japan Earthquake Task Force. 2013. Genshiryoku Saigai ni Tomonau Shoku to Nou no "Fuhyo" Mondai Taisaku toshite Kensataisei no Taikeika ni kansuru Kinkyu Teigen [Urgent Recommendations on the Establishment of Inspection Systems as a Countermeasure to the Problems of "Stigmatization" of Food and Agriculture by the Nuclear Accident]. Accessed March 14, 2016. www.scj.go.jp/ja/info/kohyo/pdf/kohyo-22-t177-2.pdf. [In Japanese]

—— 2014. *Tokyo Denryoku Fukushima Daiichi Genshiryoku Hatsuden Jiko niyoru Choki Hinansha no Kurashi to Sumai no Saiken ni kansuru Teigen* [Recommendations on the Reconstruction of Livelihood and Housing of Long-Term Evacuees of the TEPCO Fukushima Daiichi Nuclear Power Plant Accident]. Accessed March 14, 2016. www.scj.go.jp/ja/info/kohyo/pdf/kohyo-22-t140930-1.pdf. [In Japanese.]

Sheppard, Eric. 1995. "GIS and Society: Towards a Research Agenda." *Cartography and Geographic Information Systems*, 22(1): 5–16.

Soja, Edward W. 1980. "The Socio-Spatial Dialectic." *Annals of the Association of American Geographers*, 70(2): 207–25.

Sui, Daniel. 2008. "The Wikification of GIS and Its Consequences: Or Angelina Jolie's New Tattoo and the Future of GIS." *Computers, Environment and Urban Systems*, 32(1): 1–5.

TEPCO. 2015. *Tokyo Denryoku Fukushima Daiichi Genshiryoku Hatsudensho no Genjo to Kongo no Kadai nitsuite* [Current Situation of the TEPCO Fukushima Daiichi Nuclear Power Plant and Future Prospects], Tokyo: TEPCO. [In Japanese.]

Torigoe, Hiroyuki. 2014. "Life Environmentalism: A Model Developed under Environmental Degradation." *International Journal of Japanese Sociology*, 23(1): 21–31.

Torigoe, Hiroyuki, and Yukiko Kada, eds. 1984. *Mizu to Hito no Kankyoushi—Biwako Hokokusho* [Environmental History of Water and People: Lake Biwa Report], Tokyo: Ochanomizu Shobo. [In Japanese.]

Yamakawa, Mitsuo. 2013. *Gensaichi Fukko no Keizai Chirigaku* [The Economic Geography of Restoration after Disaster], Tokyo: Sakuai Shoten. [In Japanese.]

—— 2014a. "Gensaichi Fukko no Shiten" [Perspectives on the Reconstruction of Nuclear Disaster Afflicted Regions]. In *Disaster Reconstruction Support Study*, edited by FURE Support Center, 1–11. Tokyo: Hassaku Sha. [In Japanese.]

—— 2014b. "Gensai Hinansha no Kikan Iko no Henka" [Changes in the Willingness to Return among Nuclear Disaster Evacuees]. *Rekishi To Anzen*, 678: 18–31. [In Japanese.]

Yamakawa, Mitsuo, Ryota Koyama, and Hideki Ishii. 2014. "Karaki Hideaki Shi 'Fukushima Kensan Nosanbutsu no Fuhyo Higai nikansuru Nihon Gakujutsu Kaigi Kinkyu Teigen no Gimonten' eno Kaito" [Response to Hideaki Karaki's "Questions to the 'Urgent Recommendations' on the Stigmatization of Agricultural Products from Fukushima Prefecture"]. *Isotope News*, 723: 38–43. [In Japanese.]

Index

Diagrams, graphs and pictures are given in italics

Taylor & Francis eBooks

Helping you to choose the right eBooks for your Library

Add Routledge titles to your library's digital collection today. Taylor and Francis ebooks contains over 50,000 titles in the Humanities, Social Sciences, Behavioural Sciences, Built Environment and Law.

Choose from a range of subject packages or create your own!

Benefits for you

» Free MARC records
» COUNTER-compliant usage statistics
» Flexible purchase and pricing options
» All titles DRM-free.

REQUEST YOUR FREE INSTITUTIONAL TRIAL TODAY

Free Trials Available
We offer free trials to qualifying academic, corporate and government customers.

Benefits for your user

» Off-site, anytime access via Athens or referring URL
» Print or copy pages or chapters
» Full content search
» Bookmark, highlight and annotate text
» Access to thousands of pages of quality research at the click of a button.

eCollections – Choose from over 30 subject eCollections, including:

Archaeology	Language Learning
Architecture	Law
Asian Studies	Literature
Business & Management	Media & Communication
Classical Studies	Middle East Studies
Construction	Music
Creative & Media Arts	Philosophy
Criminology & Criminal Justice	Planning
Economics	Politics
Education	Psychology & Mental Health
Energy	Religion
Engineering	Security
English Language & Linguistics	Social Work
Environment & Sustainability	Sociology
Geography	Sport
Health Studies	Theatre & Performance
History	Tourism, Hospitality & Events

For more information, pricing enquiries or to order a free trial, please contact your local sales team: **www.tandfebooks.com/page/sales**

Routledge Taylor & Francis Group | The home of Routledge books

www.tandfebooks.com